W0234727

# Risk Management and Innovation in Japan, Britain and the United States

Assessing and managing risk is vitally important, and is increasingly studied, in a range of areas, including politics and international relations, finance and insurance, and innovation and valuing intangible assets such as patents and intellectual property. The degree to which innovation is encouraged or otherwise – a key factor for many businesses – depends in part on the attitude towards risk in the context in which innovation takes place. This book considers the different attitudes towards risk and innovation, and the different ways in which risk and innovation are handled in Japan, Britain and the United States. The book examines the subject broadly, within a cross-cultural and interdisciplinary context, and in detail, including discussions of risk management standards, of emerging risk, of managing risk in marketing, in the insurance industry, in banking, in patents, and in venture capital, and of how risk management in organizations has evolved.

**Ruth Taplin** is Director of the Centre for Japanese and East Asian Studies, which won Exporter of the Year in Partnership in Trading/Pathfinder for the UK in the year 2000. She received her doctorate from the London School of Economics and is the author/editor of 11 books and numerous articles. She published *Valuing Intellectual Property in Japan, Britain and the United States* with RoutledgeCurzon in 2004 and *Japanese Telecommunications: Market and policy in transition* is being published by Routledge in 2006, with the next book, *Innovation and Business Partnering in Japan, Europe and the USA*, being prepared. Professor Taplin has been Editor of the *Journal of Interdisciplinary Economics* for ten years. Currently she is a Research Fellow at Birkbeck College, University of London and the University of Leicester. She was recently appointed Visiting Professor at the School of International Business and Management, University of Warsaw, Poland.

# Routledge Studies in the Growth Economies of Asia

# Risk Management and Innovation in Japan, Britain and the United States

**Edited by
Ruth Taplin**

Routledge
Taylor & Francis Group

LONDON AND NEW YORK

First published 2005
by Routledge
2 Park Square, Milton Park, Abingdon, Oxon OX14 4RN

Simultaneously published in the USA and Canada
by Routledge
711 Third Ave, New York, NY 10017

*Routledge is an imprint of the Taylor & Francis Group*

© 2005 Editorial matter and selection, Ruth Taplin;
individual chapters, the contributors

Typeset in Times by
Florence Production Ltd, Stoodleigh, Devon

All rights reserved. No part of this book may be reprinted
or reproduced or utilized in any form or by any electronic,
mechanical, or other means, now known or hereafter invented,
including photocopying and recording, or in any information
storage or retrieval system, without permission in writing
from the publishers.

*British Library Cataloguing in Publication Data*
A catalogue record for this book is available from the British Library

*Library of Congress Cataloging in Publication Data*
Risk management and innovation in Japan, Britain and the United
States/edited by Ruth Taplin.
    p. cm. – (Growth economies in Asia series; 1)
    Includes bibliographical references and index.
1. Risk management – Japan.   2. Risk management – Great
Britain.   3. Risk management – United States.   4. Technological
innovations – Japan.   5. Technological innovations – Great
Britain.   6. Technological innovations – United States.   7.
Technology transfer – Economic aspects – Japan.   8. Technology
transfer – Economic aspects – Great Britain.   9. Technology
transfer – Economic aspects – United States.   10. Intangible
property – Japan – Management.   11. Intangible property – Great
Britain – Management.   12. Intangible property – United States –
Management.   I. Taplin, Ruth.   II. Series.
HD61.R56887 2005
368–dc22                                            2005007662

ISBN 978 0 4153 6806 3

# Contents

# Illustrations

## Figures

## Tables

# Contributors

**Gerry Dickinson** is Professor Emeritus of international insurance at the Cass Business School, City of London, and Vice Secretary General of the Geneva Association and Head of its Insurance and Finance programme. Previously he held professorships at the University of British Columbia, Vancouver, and held the CV Starr Chair in International Insurance at the American Graduate School of International Business (Thunderbird), Arizona. He serves on the International Academic Advisory Board of the IRDA's Institute of Insurance and Risk Management, Hydrabad, India, is Honorary Professor in Insurance at Wuhan University, China, and Honorary Fellow, Beijing University. He is a non-Executive Director of Trust International Insurance Company, Bahrain, and Market Insurance Brokers, London. He has authored 3 books and over 90 articles in journals and books in the field of insurance, risk management and investment.

**Matthew R. Hogg** is an Underwriter at Kiln, a managing agency at Lloyd's of London. As part of the Risk Solutions team, he is the specialist in the field of intellectual property insurance and, in particular, first-party cover. Matthew previously worked in the Lloyd's market and in Chicago for a multinational insurance broker. He regularly contributes articles to national and international trade publications, is a regular speaker at national and international conferences, and contributed to Dr Taplin's previous book, *Valuing Intellectual Property in Japan, Britain and the United States* (London: RoutledgeCurzon, 2004). Matthew has a degree in Law, a Master's degree in Law and Economics, in which he focused on intellectual property law, and holds the Advanced Diploma in Insurance. He is also an active member of the Chartered Institute of Insurance and the Licensing Executive Society.

**Masatoshi Kuratomi** is currently the Development Bank of Japan's Chief Representative in London, representing DBJ in many fields including environmental issues such as carbon finance. Since joining DBJ in 1981, he has accumulated over 20 years' experience in credit analysis on companies and projects of various Japanese industries concerning energy efficiency, urban re-development, IT and environmental protection. Previously, as Deputy Director General of Department, he supervised loans for waste management, recycling and related businesses, some through project finance. He has a Bachelor's Degree in Economics and Accounting from Kyushu University.

**Oliver Prior** commenced work as a Lloyd's Insurance Broker in 1963 with Lambert Brothers and then joined Bland Welch Ltd in 1970 and there began to specialize in insurance for financial institutions. In 1975 Sedgwick and Bland Welch merged to form Sedgwick Group (now Marsh & McLennan) and he was appointed Managing Director of the Financial Services subsidiary. In 1986 he joined Alexander & Alexander (now Aon) as a director of Alexander Howden Limited and Chairman of Holmes Johnson Lessiter Limited and Halford Shead Limited (specialist financial institution and fine art insurance brokers).

In 1990 he joined Willis Faber and Dumas (now Willis) as Chairman of the Financial Institution and Specie Division which, in 1992, was awarded the Queen's Award for Export Achievement. In 1999 he was appointed a Research and Development Director for Willis Limited in which role he served until retirement in 2004. During his time as Research and Development Director he gave over 40 external presentations on alternative risk transfer, Captive Insurance Companies etc. and developed the Willis Internal ART Training Manual and training strategy. Oliver now works as a Senior Consultant to The First City Partnership Ltd.

**Nick Schymyck** is Head of Research & Development at the Lloyd's insurer, Kiln where he is responsible for new product development, exploring innovative risk transfer solutions and research. Prior to that he was Group Insurance and Risk Manager at the media company Pearson plc where he was responsible for the company's global risk transfer requirements. Nick has over 20 years' insurance, claims and risk management experience both in the insurance market and in industry. Nick has a BA (Hons) in Economics and Politics from the University of Warwick. He is also a Fellow of the Chartered Insurance Institute and an Associate of the Institute of Risk Management.

**Ruth Taplin** is Director of the Centre for Japanese and East Asian Studies, which won Exporter of the Year in Partnership in Trading/ Pathfinder for the UK in the year 2000. She received her doctorate from the London School of Economics and is the author/editor of 11 books and numerous articles. The most recent are *Exploiting Patent Rights and a New Climate for Innovation in Japan* (London: Intellectual Property Institute, 2003); *Valuing Intellectual Property in Japan, Britain and the United States* (London: RoutledgeCurzon, 2004); and *Japanese Telecommunications: Market and policy in transition* (London: Routledge, 2006).

Ruth has been Editor of the *Journal of Interdisciplinary Economics* for ten years. Currently she is a Research Fellow at Birkbeck College, University of London and the University of Leicester. She was recently appointed Visiting Professor at the School of International Business and Management, University of Warsaw, Poland.

**Anthony Trenton** is a Solicitor-Advocate at Denton Wilde Sapte specializing in intellectual property law. He has over ten years' experience of patent litigation in a wide variety of sectors and has acted in some of the most high-profile cases in the UK, up to the House of Lords. He has a great interest in Japan and has also contributed a chapter on recent developments in patent law to the book *Exploiting Patent Rights and a New Climate for Innovation in Japan*, edited by Ruth Taplin (London: Intellectual Property Institute, 2003).

**Peter C. Young** occupies the E. W. Blanch, Sr Chair in Risk Management at the University of St Thomas College of Business. In that capacity he is responsible for the MBA Concentration in Risk and Insurance Management. Peter holds a PhD in risk management from the University of Minnesota and a Master's degree in public administration (state and local government) from the University of Nebraska-Omaha. He is considered a leading expert on risk management, particularly in public sector organizations, and has published extensively on that topic.

Peter has been a Visiting Professor to City University in London and Aoyama Gakuin University in Tokyo. Currently, he is a distinguished Honorary Professor at Glasgow Caledonian University in Scotland, and an external scholar and senior adviser at the European Institute for Risk Management in Copenhagen, Denmark.

# Preface

Similar to the current progression of intellectual property (IP), risk management (RM) is moving away from its traditional base in insurance to encompass and cross-cut many disciplines. Just as IP has become an interdisciplinary field (as elucidated in *Valuing Intellectual Property in Japan, Britain and the United States*) having moved out of the confines of accountancy, RM has become a part of legal, financial, engineering, technical and IP disciplines, not simply insurance. Such change opens the door for innovative ways in which risk can be assessed and risk-related products created. RM has always been shaped by cross-cultural influences as different cultures produce a wide variety of attitudes to risk. The literature tells us that the higher the trust in cultural institutions the less averse people are to risk taking, while the lower the trust in such institutions the more cautious they are in risk taking. Japanese culture has always been very risk averse for example, displaying a caution that has traditionally led to methodical, slow, prolonged decision-making processes in business management. With modern information technology the speed of decision-making is accelerated while the move to business assets becoming largely intangible means that the cautious approach, so valuable for business and largely comprised of tangible assets, is becoming obsolete. Japan has been forced to rapidly restructure its IP systems, as chronicled in *Exploiting Patent Rights and a New Climate for Innovation in Japan*, to once again become globally competitive. As in the rest of East Asia, risk has never been a feature of concern or academic discussion because risk was traditionally absorbed by the institutions of these societies. It is only recently, after the Asian economic crises showed these traditional institutions to be failing to absorb risk and the need to protect IP, that these countries are becoming hungry for innovative risk management solutions.

We can see very clearly how the changes brought by the upsurge in intangible assets has affected RM, which in the case of Japan can be seen clearly in the changes made to their patent courts. Technology business managers in Japan were forced to be cautious and risk averse because the patent system took too long to process patent applications and the law courts had very drawn out procedures to decide outcomes on infringement cases which, by the time of the decision, meant the technology was already obsolete. Employees' rights to compensation are also changing, a startling trend that has been happening in Japan and which has been chronicled and assessed in this series. Company-based inventors, being granted large sums of compensation money through the courts, have added a whole new meaning to managing risk in companies and a new area in which to be innovative with regards to insurance for both employers and employees. Similar issues are being discussed in England and the EU. The EU is still defining the nature of the Community Patent and the outcome will not only affect attitudes to the law in relation to patenting, but to risk in the EU countries and the UK.

The UK, EU countries and especially the US, have not been so risk averse but have been disrupted by corporate scandals and a crisis in consumer confidence that has caused many in company management to rethink their attitude to risk and to overly rapid, careless decision-making. This has led to attempts to reassess RM coupled by pressures to change fundamentally corporate governance in Japan, the UK, the EU and the US, which are all cross-affecting one another.

This book is not only needed to explain cross-cultural influences on RM, but how these different cultures are changing as a result of both external and internal pressures and are reshaping RM. The current evolution of RM needs to be reassessed in the new light of the move towards it becoming truly interdisciplinary coupled with the pressures of emerging risk, such as changing IP and terrorism, new and increased pressure from shareholders, and both governmental and non-governmental organizations such as the EU and trade associations respectively. Practitioners in this book offer practical solutions from an unparalleled experience in the field. The contributors, both practitioners and academics, have many years of experience in many diverse areas such as insurance, law, exporting, medical, publishing, international risk management etc. Many of the contributors are members of professional risk management trade associations, where they have all had daily contact with real RM problems and solutions at the cutting edge, several have lived and worked in the countries under discussion and one is a cross-cultural specialist. The wealth of such great experience imparted in these chapters can only benefit those involved or interested in RM, both conceptually and empirically, in their everyday business decisions.

Ruth Taplin

# Acknowledgements

The editor would like to thank Mr Mike Barrett, Chief Executive and Professor Peter Mathias, President of the Great Britain Sasakawa Foundation for their vision and generous support in making this book a reality. She would also like to thank all her colleagues at Kiln; Michael Faber, Chairman of Willis Japan Ltd, and Mr Masatoshi Kuratomi, Chief Representative of the Development Bank of Japan, for their valuable support of this book in terms of their time given and sponsorships; Professor Gerry Dickinson for his valuable suggestions concerning the book; Lord Levene through giving his support for the book launch; and Mr Peter Sowden of Routledge. Professor Matthias Beck, Principal Director of the Centre for Risk Management and Governance at Glasgow Caledonian University, assisted in reviewing the initial book proposal and made many welcome suggestions with regards to its contents. Finally, all the contributors must be thanked for their efforts in producing excellent material and getting their manuscripts in on time.

# 1 Introduction

## An interdisciplinary and cross-cultural approach

*Ruth Taplin and Nick Schymyck*

## Introduction

It is opportune to publish a book on risk management and innovation. It is almost axiomatic to say that risks are at their greatest during periods of change. At the moment we are seeing the rapid introduction of new technologies, for example genetically modified (GM) foods, biotechnology and nanotechnology, which bring with them exciting new opportunities. They also create a whole new set of risks, which will themselves be both a great opportunity and a challenge.

Risk and innovation are also linked to intellectual property (IP) whereby the bulk of assets in most companies are becoming intangible. The basis of IP is innovative, original ideas that make the risk greater as original ideas are a rare commodity which the less original seek to copy and infringe. In this book we cover new emerging risks and insurance responses, the increasingly complex risks associated with capital markets and new ways of protecting such risk through insurance, developments in enterprise risk management and, of course, IP – from a variety of approaches: the insurance and risk management perspective, managing risk in the Japanese and UK patent courts and then IP and bridging loans with particular reference to their emerging roles for the venture and rehabilitation of businesses in Japan. The book carries two overarching themes: first, the interrelationship between risk and innovation and, second, the need for a cross-cultural and interdisciplinary approach to risk management as set out in this introductory chapter.

## The concept of risk being interdisciplinary and cross-cultural

The concept of risk is broadening in nature and taking on an interdisciplinary nature. It is following the trend in the widening of the definition of IP into many disciplines because intangible assets have become increasingly pervasive in companies at all levels and across different professions. The definition of risk is changing as it becomes interwoven with innovation and a rapidly globalizing world. For companies to survive they must innovate at a previously unparalleled rate and in the context of greater uncertainty. This can only mean that the risks they take are deepening. The management of risk has also always varied according to the cultural context. In a globalized world it is becoming imperative to understand the variations in risk taken at a cultural level, which affect business decisions profoundly. If one society is risk averse and one risk taking it will affect the way in which business can be conducted. Further on in this chapter we explore these cross-cultural differences with regard to East Asian, North American and European cultures.

The scope of this book is, therefore, necessarily broad in its exploration of the concept of risk at a variety of levels, in different contexts and emerging considerations. Therefore, risk management is explored in different cultures, and in a variety of industries, professions and disciplines, including banking, insurance, law, accountancy, education, IP, alternative risk transfer, development banking and the patent courts.

Unsurprisingly, risk management from necessity is continuing to evolve rapidly to respond to these developments in a proactive manner. In this book we aim to move the understanding of the processes of risk to new levels.

## Definition

There are many definitions of risk management and there are many views as to where risk management as a process begins and ends and what its responsibilities should be. Some definitions are fairly narrow and some quite broad. The nature of risk reflects the fact that the future holds great uncertainty. Risks can represent both threats and dangers as well as offering significant opportunities. To fit in with the theme of innovation we have preferred to take an expansive view of risk management and this will also align itself with the advent of enterprise risk management where an organization-wide approach is taken to consider *all* of the risks facing an organization. For the purposes of this chapter we have taken risk management to mean 'the process of understanding the nature of uncertain future events and making positive plans to mitigate them where they represent threats or to take advantage of them where they represent opportunities'.[1] The definition therefore encompasses both 'upside' and 'downside' risk. The AIRMIC Risk Management Standard,[2] for example, states that 'Risk Management is increasingly recognised with both positive and negative aspects of risk'. This broad view moves risk management to the core of an organization's activities and makes it essential to the organization's strategy. Aligning risk management with upside risk may also see risk management bringing with it competitive advantage for those organizations that embed robust risk management structures. The development of enterprise risk management, holistic risk management and the Chief Risk

Officer can be seen as a natural development from a broad definition of risk management.

Above all, risk management is a dynamic process whereby intervening factors can cause outcomes to differ from those planned. Internal causes can be for the most part grouped under *operational risks*, which include fraud, system failure, the disruption of production, human error and so forth.

Other risks that will be covered in this book include emerging or new risk, systems risk, credit risk, market risk and liquidity risk. A pervasive feature of many of the risks under consideration, and particularly emerging risk, is that they are often systemic in nature – an incident in one place may have dramatic and unforeseen consequences elsewhere or in a number of different places.

Systems risk as opposed to systemic risk refers to the risk of sustaining losses that result from breakdown or malfunction of computer systems, systems defects or the improper usage of computers.

The other three risks (credit, market and liquidity risk) are considered mainly in relation to banking functions. The first is credit risk, which refers to the risk of sustaining losses resulting from a decline or complete loss in the value of assets due to the deterioration in the financial circumstances of the borrower. Credit risk management involves the monitoring of individual loans in addition to bank-wide portfolio management.

Market risk comprises interest rate risk, based on changing interest rates, and exchange risk which results from issuing foreign currency bonds and from extending foreign currency loans. Liquidity risk refers to unexpected short-term funding requirements that may occur.

## Risk and innovation

At the heart of both risk taking and innovation is the ability to adapt to change, to view change as an opportunity rather than a threat. Therefore, risk can be viewed both positively as an opportunity and negatively as a threat. With competition intensifying on international markets and research costs escalating and necessitating greater cooperation among companies in the form of innovative consortia, being overprotective of one's interests and being overcautious concerning risk taking and cooperating with other companies can only lead to poor results. The UK's Department of Trade and Industry estimates that the top 700 R&D active companies globally spent over £200bn on R&D in the financial year 2003/4. Eighty billion pounds was spent by companies based in the Americas (mostly the US), £73bn in Europe and £50bn in the rest of the world (mainly Japan).[3] At 4.2 per cent of sales for such companies R&D is clearly a massive item of expenditure.

Change is essential to the process of innovation and, in the modern world of increasing complexity and rapid change, managing efficiently the risk of changing opportunities, even if it entails innovation being created within traditional institutions in society such as the family, is the only way forward. Entrepreneurs are among the best equipped people to deal with risk as they ride on waves of change and see the majority of opportunities that come their way. Entrepreneurs recognize the abundance of opportunities that surround them and decide quickly which are the best irrespective of the risk as they think of innovative ways to manage risk, making systematically thoughtful decisions through intelligent choices, and rearranging existing resources so that they become more productive.

## Expanded role of risk management

### *Traditional role*

Until the recent past the practice of risk management has often, although not always, taken a disparate and uncoordinated approach to the risks facing an organization. This meant that risks were often treated separately with different sections of an organization taking responsibility for different risks. The issue of a silo approach to risk management has been well written about and discussed in risk management literature. Often an organization's risk manager would have been responsible for the negotiation of a firm's insurance policies and also hazard risk management, particularly in connection with the organization's physical assets such as buildings, equipment and stock. The more sophisticated or complex organizations might also buy alternative risk transfer (ART) products, a process that will be explained in Oliver Prior's chapter (Chapter 6), and these would often also have fallen within the risk manager's remit. Other risks such as legal, financial or IT would often be handled elsewhere within the organization, say by the General Counsel, Finance Director or Chief Technology Officer respectively. Thus, there was not one overarching approach to the total risks facing a firm. This disparate approach to risk also meant that some areas of risk were not even considered.

In the recent past a variety of pressures have caused risk management to broaden its approach and become future-facing to the total risks facing an organization. One overriding reason is that a traditional approach is simply no longer sufficient to match today's risks and the aspirations of organizations. At the same time a number of corporate scandals often involving the equivalent of a corporate meltdown has led to the realization that compartmentalizing or departmentalizing the risk function will simply mean that it will not be able to do its job properly. We summarize reasons for this evolution below.

### *Changing nature of business and impact of technology*

Business has become much more complex during recent decades. As mentioned above globalization continues to increase unabated. Organizations often have operations in many different territories. The use of the internet and e-commerce was minimal just ten years ago whereas now they are essential features of the way we work. The intangible assets of an organization, for example IP, reputation and brand, are now likely to form a significant proportion of a firm's total value. The types of claim an organization could face are also now far wider than ever before.

Risk management has simply evolved to meet the changing nature and requirements of business and thereby has avoided gaps appearing in the firm's approach to risk and significant issues being missed.

### Insurance and insurable risk

The period 1996–2000 was a period of plentiful insurance capacity and reducing premium rates. At that time insurers were willing to offer broad policy wordings, high policy limits and year-on-year it was usual for large companies (barring any major business expansion or significant claims activity) to see premiums fall. The fact that insurance was seemingly easy to purchase meant that insurance and insurable risk were not often very close to the top of the corporate agenda for many companies.

Towards the end of 2000 the insurance market started to harden with increasing premium rates and tightening of insurance policy and conditions. This process was sharply exacerbated by the 9/11 attacks in the US, which saw massive claims against insurers. As a result companies saw the value of their insurance covers and insurance and insurable risk rapidly moved up the corporate agenda. This was given even further impetus by difficulties, for example in the Directors and Officers liability insurance market, which saw significant claims and, subsequently, narrowing of cover and premium increases.

With companies restructuring their approaches to corporate governance emanating from Enron and related crises, it is becoming increasingly important that intangible assets, goodwill, reputation and the intellectual capital of companies are protected. Insurance is beginning to offer potential solutions, but cannot be considered a wholesale replacement to comprehensive risk management due to the potential magnitude of loss.

### Corporate governance

Probably the main reason for the expansion of the risk management role has been the impact of corporate governance. A significant driver behind this has been the number of corporate and accounting failures and scandals over recent years leading to pressure from shareholders to tighten control and risk management. We are all aware of Maxwell, Enron, Barings Bank, WorldCom and Global Crossing from the recent past. There has also been a growing realization that taking a reactive approach to risk or splitting risk into silos is simply not good enough and is potentially dangerous in that no overall view of risk is taken. Thus, form filling and box ticking are out and a wide-ranging proactive and top-down approach is definitely in.

The impetus in the UK from corporate governance to take a coordinated approach to risk management can be traced back to the 1990s with the publication of the Cadbury Report (The Financial Aspects of Corporate Governance), Greenbury Report (Directors' Remuneration) and Hampel Report. These were drawn together by the Hampel Committee on Corporate Governance with the issuance in June 1998 of the 'Combined Code: Principles of Good Governance and Code of Best Practice'. As a further development, guidance to assist UK listed companies to implement the Combined Code in relation to internal control was provided in 1999 by the publication of 'Internal Control: Guidance for Directors of the Combined Code' by the Internal Control Working Party of the Institute of Chartered Accountants in England and Wales (chaired by Nigel Turnbull). This is also known as the Turnbull Report.[4] The importance of the Turnbull Report should not be underestimated for the practice of risk management and it's subsequent development.

The most significant feature is the adoption of a risk-based system of internal control: 'The guidance is based on the adoption by a company's board of a risk-based approach to establishing a sound system of internal control and reviewing its recommendations' (Paragraph 9).[5]

Also: 'A company's system of internal control has a key role in the management of risks that are significant to the fulfilment of its business objectives' (Paragraph 10).[6]

Other significant features include an annual (i.e. regular) review:

> The directors should, at least annually, conduct a review of the effectiveness of the group's system of internal control and should report to shareholders that they have done so. The review should cover all controls, including financial, operational and compliance controls and risk management.
>
> (Paragraph 3)[7]

The Report goes on to point out that the risks facing a company are 'continually changing' and that a sound system of control depends on a 'thorough and regular evaluation of the nature and extent of the risks to which the company is exposed' (Paragraph 13).[8]

Paragraph 13 points out that profits are a reward for risk taking and therefore the purpose of internal control 'is to help manage and control risk appropriately rather than eliminate it'.[9] One of the major features of innovation will always be 'risk', and therefore risk management needs to facilitate innovation rather than stifle it.

It is interesting that the Turnbull Report sees risk management in its dynamic context. In the context of emerging risk and the theme of innovation of this book it should be noted that the Turnbull Report states that the system should 'be capable of responding quickly to evolving risks to the business arising from factors within the Company and to changes in the business environment' (Paragraph 22).[10]

Finally, the Report sees that the system of internal control should 'be embedded in the operations of the company and form part of its culture' (Paragraph 22).[11] Therefore, it is clearly not appropriate to assign responsibility for risk management to one person or group within an organization.

The Turnbull Report therefore envisages a dynamic top-down approach to risk management, embedded within an organization, encompassing the wide range of risks that an organization faces and with regular reporting.

## Subsequent developments

### *In the UK*

Since the Turnbull Report, the Combined Code has continued to evolve with the issuance by the UK's Financial Reporting Coucil of a revised Combined Code in July 2003. This included recommendations from the Higgs Report,[12] for example, that at least half of the Board of a listed company should comprise independent non-executive directors and that the Chief Executive should not go on to become a Chairman of the same company. It also included the Smith guidance on audit committees.

### *Elsewhere*

Outside the UK we have seen developments such as the Sarbanes-Oxley Act (2002) in the US which sets out clear rules for corporate governance and, in particular, certification of financial statements by the CEO and CFO of a company and also ensures the independence of auditors and the audit committee.

'Basel II', which is relevant to the banking industry and is due to be implemented in 2006, is an accord aimed to align regulatory capital more accurately with operational and credit market risks that international banks face. The insurance industry has its own equivalent to 'Basel II', with Solvency 2. Issues relating to this will be dealt with in Oliver Prior's and Gerry Dickinson's chapters, 5 and 6 and 8 respectively.

## Risk management approach

A key theme of the Turnbull Report is that risk management should be embedded within an organization. In a sense, therefore, risk management becomes everyone's responsibility and is a reflection of 'good management'. The Risk Manager or Chief Risk Officer therefore becomes a facilitator assisting the organization to set up processes to identify, assess and control risk and ensuring the smooth (and regular/ongoing) operation of these processes.

## Risk identification

The start of the process is to identify and classify the risks facing an organization. The process may initially involve the identification of all risks facing the organization or just a selection of the main risks. If the approach is to identify *all* risks a process needs to be introduced to screen the risks and eliminate those that are relatively insignificant. At the end of the identification process it should be possible to have a group of significant risks to consider.

The challenge in connection with new or emerging risks arising from, for example, technology or innovation is to identify whether such risks are significant when either minimal or no historic data are available. It may be that the cautionary approach will be to include these and monitor them.

There are various means by which the significant risks for an organization can be identified – often a process that requires some interaction between key personnel within an organization, e.g. facilitated brainstorming may be one of the better means of bringing out a wide selection of key risks. Other methods might include discussion or debate of key risks, risk reviews and specific studies, surveys, structured interviews, management reports and checklists/questionnaires. The key message is that this should be an internal process and external assistance is better placed to facilitate rather than carry out the identification process. Ideally the process should be as interactive as possible.

In any event any identification process is likely to throw out a wide variety of risks. It should be noted that it is important that a broad spectrum of participants should take part in any identification process.

Once identified, the risks should be grouped:

- financial – currency fluctuations, credit, interest rates, pension fund;
- hazard (i.e. traditionally insurable) – property risks, natural catastrophe, terrorist attack, business interruption, product liability, employee injury;
- legal/compliance – litigation risk, errors and omissions, patent infringement, breach of regulations/laws;
- operational – IT, outsourcing, logistics, interdependencies;
- people – succession planning, key individual, training;
- strategic – customer preferences/taste, market share, technology, R&D, brand, reputation, emerging markets.

Some of the risks may also overlap into other areas. Outsourcing, for example, is an operational risk but there are also other aspects, for example, people. An employee injury is likely to arise from a hazard risk but there will also be legal risk issues, for example, potentially very significant litigation.

In this book we explore a wide range of variables both external and internal that impinge on such risk taking.

*Assessment*

Once risks have been identified it is necessary to assess and prioritize each risk. Measurement of risk can be problematic because quantitative data and historic data may not always be available or may not be particularly good. Also, while certain types of risk, such as hazard, may be relatively straightforward to quantify, there are others, such as strategic risks, which simply do not lend themselves to quantification.

Often a risk matrix approach is used (Figure 1.1).

It is critically important to prioritize the risks to decide which are important and the action that should be taken in respect of each. This might be achieved as follows.

Prioritization:

1  High severity/High probability – top priority i.e. take immediate action.
2  High severity/Low probability – major priority i.e. consider action – probably have a contingency plan (depending on the risk).
3  Low severity/High probability – lesser priority i.e. consider action (contingency plan probably not necessary as severity is low).
4  Low severity/Low probability – low priority i.e. take no action but monitor in case the nature of this risk should change.

Other organizations might use a traffic light formulation:

1  Red – Take immediate action.
2  Amber – Consider action – possibly have a contingency plan.
3  Green – Take no action but monitor.

*Risk control*

Once risks have been identified and measured/prioritized it is necessary for the organization to decide its control strategy. This is the whole purpose of risk management.

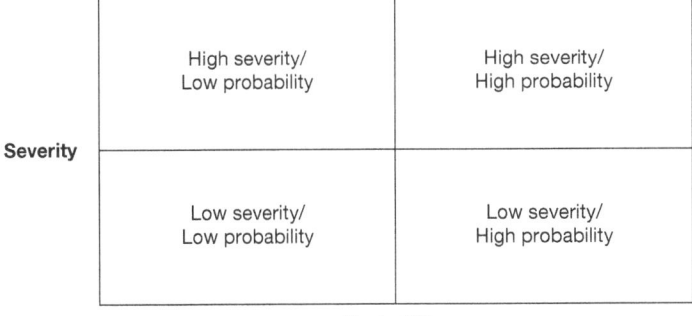

Figure 1.1 Risk matrix

Control strategies could consist of:

- Acceptance – the risks may be both 'low severity' and 'low probability' and any potential impact of the risk (if it happens in the first place) may be small; in addition, as mentioned risk is often the price of innovation – provided the risk has been considered the organization may simply decide that the risk is one it can live with.
- Transferring the risk – either through the purchase of insurance, which is an approach we are all familiar with, or, alternatively, the risk may be transferred to another organization in contract.
- Control – this could be achieved by introducing better or more stringent operational control or more frequent audits. It is interesting to note the expansion of the internal audit and risk management functions in recent years. In some organizations new control functions have been created, often involving the merger of internal audit with risk management.
- Risk sharing – sharing the risk, for example, by forming a Joint Venture company either in the development of a new product or when deciding to enter a new market – this approach plays to each participant's strengths and should minimize risks for each partner.
- Avoidance – in extreme circumstances risks should be avoided but we should bear in mind the caveat that profit, by its very nature, brings with it risk and companies are in the business to make a profit. An overcautious approach is likely to lead to sub-optimal profits.

Two further factors should be remembered:

- the approach to each risk must fit in with the strategy of the organization;
- a person or specific group should have responsibility (ownership) of the risk to ensure that any action or monitoring is carried out – where a risk is everyone's property it is soon likely to become no one's property.

## The future of risk management

### Impact of corporate governance and regulation

This is set to increase incrementally in line with business developments, shareholder demands and increasing demands from regulators. The Turnbull Report was, at the time, very far reaching and it is likely that further developments will simply extend Turnbull further, for example, the Higgs Report. Other countries are taking their own steps on corporate governance and risk management, e.g. the Sarbanes-Oxley Act in the US and the Dey Report in Canada.

At the same time there have been murmurings within business that corporate governance and regulation are going too far and distracting executives from running their businesses. For example, there were several reports in the business press in late 2004 that certain European companies might consider de-listing from stock exchanges in the US due to the costs of compliance with the Sarbanes-Oxley Act. The situation continues to develop and it is too early to predict whether these murmurings will turn into anything concrete.

### Enterprise risk management

It seems that enterprise risk management (ERM) is here to stay and it is likely that as the benefits are demonstrated ERM will continue to extend beyond listed companies into private companies and government organizations. There seem to be growing pressures, for example, for private equity to recognize recent developments in corporate governance and become more accountable and transparent.[13] The danger is that the corporate governance and risk management models used, for example, by public companies may not be applicable to private equity and may actually take away many of its advantages. Gerry Dickinson (Chapter 8) deals with the recent trends in ERM.

### Chief Risk Officer/Risk Manager

With the Turnbull Report and ERM the Risk Manager now has a broader remit than before. The result is that while risk management continues to attract insurance professionals it also now attracts people with other qualifications and specialized skills such as accountants, lawyers and so forth. Anyone who reads the business press will notice that the market for risk management and compliance professionals is burgeoning – particularly in the financial services industries.

### Emerging risk

Emerging risk is likely to become a major feature of the twenty-first-century risk environment and Risk Managers/Chief Risk Officers will need to understand and grapple with major new risk scenarios from innovations and new technologies.

The way we work also continues to evolve: for example, the tendency to outsource key processes and now the move to take this a stage further by offshoring. These add the risks of entrusting key processes to a third party and also other features such as political risk when the processes are moved to a different territory. Particularly where any of these processes are customer-facing an organization may also be adding the critical element of reputational risk. In addition, the world economy continues to globalize and certain economies, notably China, are growing rapidly. Mergers and acquisitions activity, though much reduced from several years ago, continues to be a significant feature.

We are also in a period where we are beginning to encounter noticeable climate change and are witnessing an increase in the frequency and severity of natural catastrophes. In 2002 many parts of Europe suffered serious flooding but in 2003 there was a record-breaking heatwave. There was also the tragic Indian Ocean earthquake and subsequent tsunami at the end of 2004, which caused massive loss of life, significant property damage and economic dislocation to countries around the Indian Ocean. That year also saw four hurricanes (Charley, Frances, Ivan and Jeanne) strike the southern US and various parts of the Caribbean. The costs to the insurance industry of these hurricanes may even exceed the costs of insurance claims arising from Hurricane Andrew in 1992. Inclement weather and earthquakes in Japan as well as unauthorized logging in the Philippines are adding new dimensions to risk assessment. For example, 2004 also saw typhoons Chaba, Songda and Tokage in Japan. Systemic risk is developing into a significant feature of the risk environment as a reflection of the sheer interconnectedness of twenty-first-century life. Finally, of course, there has been the massive impact to risk management following the 9/11 attacks in the US.

It will be crucially important for risk professionals to have both a flexible and 'outside the box' approach to new risks where our knowledge database of such risks is likely to be severely limited. Nick Schymyck reviews emerging risk and developing responses in Chapter 2.

In Chapters 5 and 6 Oliver Prior notes the changing role of the insurance intermediary in response to new and emerging risks. He notes that insurance transactions tend to divide into three types:

- One-dimensional – these are insurances where the policy language is governed by regulation and is largely standardized, e.g. employer's liability or motor insurance. Here the only negotiable items between the seller (the insurer) and the buyer (either a company or individual) are limits of indemnity, levels of self-insurance and (to a limited extent) premium. Many insurance products that fall under this heading are commoditized and the customer decision whether to insure or not with a particular company is often largely determined by price.
- Two-dimensional – this is where the cover can be described in a template insurance contract but comprehensive alteration is often required to adapt the cover provided by the template contract to meet the specific needs of a particular insurance buyer. This usually results in a manuscript policy being produced. In addition to the contract document, limits of indemnity, levels of self-insurance and premium are negotiable items.

•   Three-dimensional – this is where no template insurance contract exists
    and it is necessary to construct a completely new insurance policy or
    take an alternative approach (e.g. alternative risk transfer) to meet the
    specific needs of an insurance buyer. In addition to the contract docu-
    ment, limits of indemnity, levels of self-insurance and premium are
    negotiable to a certain extent.[14]

It is the two- and three-dimensional approaches to risks that are considered in this book.

## Insurance, risk management, innovation and intellectual property

Billions of dollars a year are spent by industry on R&D in order to
generate the next product or service success story. Such innovation needs
to produce more than simply a short-term cash-cow to recoup R&D expenditure, and the reality is that innovation today produces intangible assets:
a much wider concept. These assets will preferably imbue ownership in
their creator with property rights allowing exploitation of their developments to the exclusion of others in order to achieve a suitable return on
R&D. Today around 75 per cent of business assets are intangible assets,
subject not only to the risk of 'creative destruction' described by economist Joseph Schumpeter[15] but also a number of other perils both insurable
and uninsurable. These perils are often grossly neglected by businesses
who, in line with tradition, continue to protect their often less valuable
tangible assets such as plant and stock. Intangible assets bring both great
risk and great reward but such fruits of innovation must be greater understood, managed and protected. Chapter 3 by Matthew Hogg will explore
current attitudes, shortcomings and the potential of third-party risk transfer,
commenting that there must be a greater interdisciplinary approach to
intangible asset protection to incorporate the entire intellectual capital of
the organization.

Changes linked to risk, innovation and IP are also permeating the institutional structures of our societies. This can be seen as outlined by Anthony
Trenton in Chapter 4 in relation to the patent courts. Japan and England
are attempting to make the patent procedures less risky by speeding them
up and reducing cost but not at the expense of quality investigations and
judgments. In Japan in particular, the process involved in patenting took
so long that the new technology was out of date before it received a patent,
which also dampened enthusiasm for innovation. As Chapter 4 explains,
enforcement procedures against infringers are being made fast, accessible
and specialist, which also reduces risk as it becomes clear those who are
infringing will not be able to get away with their unlawful damaging
actions. The outcome of the discussion regarding the Community Patent

and European Patent Litigation Agreement, also discussed in Chapter 4, will serve to define the type of risk and attitude towards support for innovation by the European Community.

## Patterns of cross-cultural risk management

In comparing management styles and decision-making methodically across a number of countries we can see tendencies towards risk taking and how the process of globalization is putting pressure on different societies to change their attitude to risk out of necessity. Analyses in relation to risk and different cultures have been few and far between, mainly emanating from the sociological work of Aaron Wildavsky, in 'Choosing preferences by constructing institutions – a cultural theory of preference formation',[16] or Mary Douglas in 'Witchcraft and leprosy: two strategies of exclusion'[17] and the economic/management analysis of Ruth Taplin, 'The new Silk Road – amalgamating Western and Asian style management'.[18] Wildavsky argues that risk is one of the areas that most strongly reflects cultural predisposition. In the case of business it can be argued that, when operating in a globalized environment, culture most strongly predisposes attitude to risk and management decision-making. Wildavsky and Douglas discuss further the idea that those people who have the most trust in their institutions are risk accepting and those that have low trust in their institutions are risk averse.

However, culture is an essential, often institutional, variable that affects how management decisions are made and what degree of risk is to be taken. Business communication can only benefit from clarifying the degree of risk a business partner is willing to take and how this affects management decision-making. A risk averse culture such as Japan or Germany will produce great caution, and an aversion to taking high levels of risk that is not conducive to the promotion of entrepreneurship. Other cultures produce managers that are willing to take high risks, such as America or Korea, or Finland or mainland China, and that are willing to innovate and support high degrees of entrepreneurship. Then there are many shades in between, including such countries as Britain, France or Italy that exercise risk taking and the promotion of innovation, but show more caution such as the case of Britain or tend towards supporting high risk as in Italy. Risk management, and decision-making and innovation also function in different societies through a variety of institutions. In China, India and Italy for example, the management of risk by companies is exercised through the institution of the family. With family support and resources, risk taking can be high and decisions flexible, which promotes innovation and entrepreneurship. In the case of the family, decision-making is flexible and consensus is easily reached because the decision-making process is contained in one of the oldest institutions, that of blood-related individuals where the trust is high and, in the case of patrilineal societies, where the male autocratic head makes most of the decisions, which expedites decision-making.

Chen Min, in *Asian Management Systems*,[19] noted that those who follow the Confucian ethics have a collective business advantage in that they are able to mobilize more easily their network of relatives, trusted friends and business contacts. This is especially true for China, countries that have a preponderance of Chinese citizens such as Singapore or those with large segments of Chinese traders. In India, which is a country of mainly practising Hindus, networks of relatives pool their resources and manage risk flexibly through the family. The Chinese family structure is more authoritarian and less prone to splintering from in-fighting among inheritors. The Chinese have always practised primogeniture where the eldest son is the main inheritor, while in Indian culture inheritance is split between siblings, which allows for greater conflict in decision-making. Such conflict can inhibit risk taking while decision-making by one senior person from the top down allows for rapid, flexible decision-making and risk taking. If a process of consensus has to be reached as in the Japanese system, where company management controls decision-making, speed of decision-making is retarded. The greater number of people involved, the more the element of risk becomes gradually eroded with each doubtful input that does not take ultimate responsibility. The Japanese system, with its devolved sense of social responsibility and consensus, which leads to slow decision-making, lends itself to a risk averse management style. If we look at differing management variables such as: control vs. delegation; individualism, collective teamwork and invention; management by objectives vs. management by consensus; and the top-down approach vs. the bottom-up approach, the outcome of assessing all the variables can provide direction to the type of risk taking that prevails in a particular culture. If we look at three different cultures of East Asia that have all been influenced by Confucianism we can see how different management styles have led to particular forms of risk taking. See below for an illustration of this process.

### Control vs. delegation

1　Japanese – control in the company is delegated and not held only by senior officials.
2　Korean – companies are controlled by their senior officials.
3　Chinese – control is usually held by senior members of the company, but delegation can occur if it aids successful business decisions and is kept within the family.

### Individualism, teamwork and invention

1　Japanese – individualism is discouraged in favour of teamwork. Japanese people feel loyalty and pride in relation to their companies, sections and even to government. Invention is discouraged if it interferes with teamwork although recent high awards given to inventors in employee's rights to compensation suits may change this to a degree.[20]

2   Korean – individualism in the Western sense is not encouraged. Teamwork, loyalty and pride in relation to their companies and government is high. Korean people will make tremendous efforts and work together if encouraged by the Chairman of their organizations. Koreans are more inclined to take higher risks than Japanese counterparts, especially with regard to being inventive and creative, because responsibility for risk taking lies ultimately with the Chairman who can always be blamed for the actions of underlings.
3   Chinese – managers tend to be entrepreneurial and do whatever works. Therefore, loyalty tends to be to new ideas that facilitate inventions, rather than to organizations or government. Traditions can be quickly adapted to suit the needs of a niche market.

All these cultures have in common respect for senior officials. Job security is valued and younger members of a company do not expect to jump promotional ladders to senior positions rapidly. Confucianist ideas of paternalism prevail and seniority of ranking and hierarchy is reflected in the language.

## Comparative management styles

### Management objectives vs. management by consensus

1   Japanese – consensus has to be established by the group before objectives are finalized.
2   Korean – the Chairman is responsible for setting objectives and through the charisma and position of the Chairman, the employees follow the leader's objectives.
3   Chinese – this is in between by objectives and by consensus. The senior relative obtains objectives autocratically, but involves family members/employees in decision implementation to support rapidity of decision-making.

### Top-down approach vs. bottom-up approach

1   Japanese – both approaches apply as the amount of discussion for decision-making among all levels of managers is very high. Middle management is important for information gathering from the company to reach consensus. Ideas and commitment are established from the bottom upwards for ratification at the top.
2   Korean – Korean management takes a top-down approach as it tends to be authoritarian in nature with a vertical hierarchy. Middle managers are quite powerless in the decision-making process.
3   Chinese – decision-making is top-down, made by the family head or senior head manager. Management discussion is flexible, however,

and can involve consensus decisions depending on the situation and the need to meet market requirements quickly.

Both Korean and Chinese people are more accepting of risk taking because they trust in their institutions of paternally run companies whereby the Chinese trust as well in their smaller institution of the family and in the decisions of the family head.

In the Japanese case, trust has to be spread among many counterparts as decision-making is delegated. Fierce rivalry is also tolerated if not encouraged between rival intra-company teams and groups. This militates against trust and contributes to risk averse attitudes.

Until recently, concepts of risk, let alone risk management, were never a consideration in Japanese, Korean and Chinese companies. It is only recently with the increasing importance of issues of IP that risk has even been seen as a viable issue that requires attention.

## Family-based companies and risk absorption

Like consultancy, which has been viewed traditionally in East Asian countries as something that is provided gratis by family and friends (who are regarded as extended family members), anything which incurred risk was absorbed into the family business or into the corporate business. In the Japanese case, risk had the safety net of a large corporation that had many branches and cross-shareholdings, both in the traditional family owned *Zaibatsu* and modern day *Sogo Shosha/Keiretsu* (trading arms). The *Zaibatsu* were the traditional family-owned large companies that have dominated Japan from the sixteenth century beginning with Mitsui *Zaibatsu* founded by the Mitsui family who intermarried with the noble lords (*daimyo*) becoming the bankers to them and the Japanese rulers of government. These companies owned everything from a bank that was central to their organization, to coal mines, food shops, textiles, exporting/trading companies and so forth. The businesses were all interrelated and were treated as a family enterprise despite their large size. All the shares were cross-shareholdings. After the Second World War and the defeat of Japan, the Americans instituted the Spencer-Day Agreement, which introduced anti-trust law and broke the family hold on the *Zaibatsu*. This led to the reformation of the *Zaibatsu* but as non-family owned conglomerates known as *Sogo Shosha* with *Keiretsu* that continued with cross-shareholdings.

The traditional Korean *chaebols* have operated in much the same way as the old Japanese *Zaibatsu* with paternally related brothers operating the conglomerate businesses. This has changed recently with the *chaebols* that were founded in the 1920s and 1930s now being required to have outside directors filling half the places on their Boards of Directors. Samsung, the largest of the conglomerate groups in South Korea, is now 55.4 per cent and 60 per cent foreign owned with reference to Samsung Group and the overall Samsung parent company respectively. Korean *chaebols* as well as *Sogo Shosha* have many interconnected parts of their business where risk continues to be absorbed by shifting any problems to another branch of the company despite the company structure no longer being family dominated. Failed enterprises in both the Japanese and Korean cases were traditionally hidden by being absorbed by another branch of the large interlinked companies.

In the Chinese case risk has been traditionally absorbed by the family resources which are pooled together. Therefore, risk was never viewed as a process separate to the overall workings of the companies. Even in regard to banks, which, for example in the Japanese case, were part of the large conglomerate companies, borrowing to avoid bankruptcies and to avoid risk was made easier by being part of a larger organization that was backed up by cross-shareholdings and other successful, strong, branches that held reserves of cash that could be drawn upon.

It was not until the Asian crisis in the late 1990s that Japanese, Korean and Chinese businesses were made to experience the vulnerability of not having the ever-present traditional safety nets to absorb and cushion risk as many companies were close to bankruptcy. This was brought home even more intensely in the Korean case by foreign banks buying and owning native banks. The Koreans as a nation moved quickly to avoid total collapse of the economy by donating jewellery to the national coffers to build up capital reserves acting as a national absorption unit for risk and for managing the risk.[21]

The Japanese are becoming more interested in a more formalized risk management system because they realize the need to protect their IP to become once more globally competitive and they are being forced to seek inward investment from the outside world which attacks the security of being able to absorb all risk eventually in Japan Incorporated and opens up the Japanese companies and mentality to the idea of having to take risk that is dependent on foreigners. Showing a more proactive stance in protecting its IP, Japan's anti-monopoly watchdog gave Microsoft a warning in July this year regarding a contract clause for electronics companies that Toshihiro Hara, an official with the Fair Trading Commission, said amounted to preventing Japanese computer makers from demanding damages or royalty fees, even when rivals violated patents for important technology. The warning was the first in the world. Hara was concerned that the clause would discourage motivation to develop audio-visual technology and may hinder competition in the high-tech field.[22]

In its annual White Paper, the Ministry for Economics, Trade and Industry (METI) encouraged active protection of all intellectual assets in Japan, stating that efforts to offer unique goods and services have become more important than price competition. The document was approved by the cabinet in July 2004.[23]

This means that, similar to what happened in the nineteenth century with the adoption of German technocratic law, because Japan needed new institutions and laws to deal with the outside world, risk management, like IP, is adopting non-Japanese constructs to deal with issues of risk that cannot be dealt with internally to Japan.[24]

There are examples, however, where Japan is using its own internal institutions to deal with risk, in this instance the risk of re-starting a bankrupt business, allowing a Japanese governmental institution, the Development Bank of Japan (DBJ) to absorb the medium to high risk. Masatoshi Kuratomi describes in detail in Chapter 9 how the DBJ is assisting SME companies in particular to manage the risk of re-starting bankrupt businesses using IP as bridging loans and for leveraging purposes. DBJ is also encouraging risk taking through innovation by supporting new business through financial and advice packages or a new direction of an existing business and, again, using IP for leveraging purposes.

This book provides fresh insight into the meaning of risk management in taking risk and emerging risk to new levels of analysis involving an interdisciplinary, cross-cultural approach moving the entire debate forward.

## Notes

1 See 'Risk Management: a guide to good practice', The Chartered Institute of Management Accountants, 2002.
2 See 'A Risk Management Standard', AIRMIC, *ALARM*, IRM, 2002.
3 See *The 2004 R&D Scoreboard: the top 700 UK and 700 International companies by R & D Investment*, The Department of Trade and Industry, 2004.
4 See *Internal Control: guidance for directors on the Combined Code*, Institute of Chartered Accountants in England & Wales, 1999.
5 See *Internal Control: guidance for directors on the Combined Code*, Institute of Chartered Accountants in England & Wales, 1999 – Paragraph 9/page 4.
6 See *Internal Control: guidance for directors on the Combined Code*, Institute of Chartered Accountants in England & Wales, 1999 – Paragraph 10/page 4.
7 See *Internal Control: guidance for directors on the Combined Code*, Institute of Chartered Accountants in England & Wales, 1999 – Paragraph 3/page 3 (referring to Provision D.2.1 – Combined Code of the Committee on Corporate Governance).
8 See *Internal Control: guidance for directors on the Combined Code*, Institute of Chartered Accountants in England & Wales, 1999 – Paragraph 13/page 5.
9 See *Internal Control: guidance for directors on the Combined Code*, Institute of Chartered Accountants in England & Wales, 1999 – Paragraph 13/page 5.
10 See *Internal Control: guidance for directors on the Combined Code*, Institute of Chartered Accountants in England & Wales, 1999 – Paragraph 13/page 5.
11 See *Internal Control: guidance for directors on the Combined Code*, Institute of Chartered Accountants in England & Wales, 1999 – Paragraph 13/page 5.
12 See *Review of the Role and Effectiveness of Non-executive Directors* by Derek Higgs, Department of Trade and Industry, 2003.
13 See article 'Draw a veil' in 'Survey: private equity', *The Economist*, 25 November 2004 and www.economist.com.

14   Oliver Prior also notes that this process could lead to a clear demarcation, as far as intermediaries are concerned, between transaction only services and consulting and advice services relating to the transaction. This is not unlike the process that occurred in stockbroking a few years ago where the execution only broker became a feature of the landscape.

The question facing many insurance buyers is – if the intermediaries become more 'execution only' but the need for advice and consulting services increases, as more and more insurances become two- and three-dimensional, then where can such advice be found?

15   Joseph Schumpeter, *History of Economic Analysis. Capitalism, socialism and Democracy* (New York: Harper Torchbook Edition, 1976).

16   Aaron Wildavsky, 'Choosing preferences by constructing institutions – a cultural theory of preference formation', *American Political Science Review* Vol. 81 No. 1, March 1987, pp. 3–22.

17   Mary Douglas, 'Witchcraft and leprosy: two strategies of exclusion', *Man: New Series* Vol. 26, No. 4, Dec. 1999, pp. 723–736.

18   Ruth Taplin, 'The new Silk Road, amalgamating Western and Asian style management', Presentation to Managing Asia Pacific Business into the Twenty-first Century, 6–8 November 1995, Westin Stamford International Management Conference (Singapore).

19   Chen Min, *Asian Management Systems* (London: Routledge, 1995).

20   The issue of employees' rights to compensation is covered through the series of books initiated and edited by Ruth Taplin. For a whole chapter devoted to this issue see 'Employee's rights to invention – a changing situation', by leading expert Prof. Katsuya Tamai in Ruth Taplin ed., *Exploiting Patent Rights and a New Climate for Innovation* (London: IPI, 2003) and for further discussion in Ruth Taplin ed., *Valuing Intellectual Property in Japan, Britain and the United States* (London: RoutledgeCurzon, 2004). In this book see Matthew Hogg's Chapter 3.

21   All the above on cross-cultural analysis in East Asia is based on original work by Prof. Ruth Taplin in the course of lectures, briefings and material mentioned in note 17 above.

22   'Japan warns Microsoft over patents', *International Herald Tribune*, Wednesday 14 July 2004.

23   *Big News Network*, 3 July 2004.

24   See 'Overview: Japanese attitudes to litigation and IPR', by Ruth Taplin in Ruth Taplin ed., *Exploiting Patent Rights and a New Climate for Innovation in Japan* (London: IPI, 2003).

# 2  Emerging risk

*Nick Schymyck*

## Introduction

During the past few years many of our previous certainties have disappeared. It is difficult to pinpoint exactly when this process began, although it is noticeable that particularly since the 1980s the rate of change in many different spheres has increased dramatically and continues to move apace. Since that time the Berlin Wall has gone, the Soviet Union collapsed, the Cold War ended and some previous world views and assumptions now appear tired or irrelevant. New opportunities have appeared: globalization and increases in economic growth in many countries, the eastward expansion of the EU and the growth and widespread use of the internet. With the end of the Cold War old fears have been replaced with new ones which have become much more real since the 9/11 terrorist attacks in the US.

The appearance of new opportunities has coincided with the appearance of new or emerging risks. This can only be expected. The invention of the railway locomotive brought a significant dynamic to the Industrial Revolution. It created vast new possibilities such as making cheap, rapid travel available to all but the very poor; it also opened up new territories and markets such as the Mid-West in the US. It also brought with it new risks and unintended consequences such as the injuries and deaths suffered by the navvies who built the railways. The aircraft brought tremendous opportunities such as the ability to traverse continents in just a few hours and made the world much smaller. Much more sinisterly, it also introduced far more deadly and destructive forms of warfare (and more recently terrorism) than ever before.

Therefore, new technologies and new opportunities create new risks. This is exacerbated due to the rapid pace with which such new technologies are introduced. We can see this with the significant potentials of new technologies such as biotechnology and nanotechnology, but these also bring with them significant risks which we are only beginning to understand. Over time, however, these new risks often diminish as our knowledge of them increases along with our ability to manage them. New technologies, by their very nature, are full of unknowns. The danger is that we become ultra-cautious and stifle everything from innovation to entrepreneurship, and this would go against the whole theme of enterprise risk management i.e. of embracing both downside and upside risk. There needs to be a balanced approach to reflect both the opportunities and risks of innovation. In addition, when there are so many unknowns with regard to the risks of a new technology are we talking about anything real or tangible in the first place? Perhaps when the state of knowledge around a new technology improves then we may realize that the risks weren't real i.e. that we were looking at a phantom rather than a real risk.[1]

Emerging risk has recently entered the vocabulary of insurance underwriters, brokers and risk managers. Emerging risk literature has started to appear and discussion groups of concerned risk managers established, for example, the UK's Institute of Risk Management Emergent Risks Special Interest Group.[2] These developments represent a widespread desire in the insurance industry and among risk managers to understand both the unknown and also the dynamics of a rapidly changing risk environment.

There seem to be three processes taking place:

1   Previously unheard-of risks are appearing and the challenge is to identify, understand, analyse and ultimately (where possible) manage these risks.

2   The nature of some existing risks seems to be changing. For example, weather-related losses are nothing new but there seems to be evidence of climate change exacerbating weather patterns and causing much more severe and unpredictable weather losses, such as the serious floods in parts of Europe in 2002 and then the European heatwave in 2003. The film 'The Day After Tomorrow' may be a little extreme, but something definitely appears to be happening.

3   There is an increased pervasiveness of certain types of risk and we can see this in certain features of the way we live that can significantly affect the spread of a major loss event. For example, the interconnectedness of early twenty-first-century life, the interrelation and mutual dependencies between businesses, population growth and the increased concentration of people into urban areas and appearance of megacities, and the ability to travel around the world rapidly. All of these help to contribute to systemic elements of many emerging risks and assist with rapid spread when something goes wrong. There are various examples, such as the rapid spread of computer viruses in spite of rapidly improving technologies to counter these, for example, the widespread use of firewalls in computer networks. In August 2003 a minor occurrence in Ohio caused an electricity blackout covering much of the north-eastern US and Canada. The resultant disruption was massive and economic losses amounted to several billion dollars. The SARS outbreak in 2003 rapidly spread from China and Vietnam to Canada and the US. Insurers and risk managers are well aware of the issues arising from global supply chains and extensive interdependencies between businesses; a fire at a factory in one continent, for example, can soon halt production at a facility thousands of miles away.

It is critical that the insurance industry gains an understanding of emerging risk for several reasons.

First, it is essential that insurers fully understand and adequately price the risks of the insurance business that they underwrite. In the context of emerging risk is there any historic loss experience available? Is it possible to model the risks of loss? Are there any (or many) unknowns? Is there the potential for claims to appear many years in the future i.e. is there likely to be a significant claims tail with new claims appearing in a few years time. Here it is worthwhile to recognize that legal frameworks are becoming more expansive, often in favour of the claimant and with the imposition at times of strict liability regimes on new technologies. There will thus always be some uncertainty surrounding the future claims environment for casualty or liability classes of insurance business and this increases in the context of new technologies.

Second, to be able to do its job properly and to provide its customers with the products it needs it is incumbent on insurers to understand emerging risk and to continue to respond to the needs of business in an imaginative and innovative but considered way. After all ultimately the purpose of the insurance industry is to serve business in its risk transfer and risk management requirements.

Finally, emerging risk can represent a significant business opportunity to the insurance industry. For example, cyber-risk insurance products were practically unheard of a decade ago but are now entering the mainstream of insurance products required by business. We can also think of the growth of alternative risk transfer (ART) products aimed at the natural catastrophe market. For insurers, emerging risk therefore represents a significant risk to their own businesses but, if handled correctly, may also represent a significant business opportunity. In the context of enterprise risk management risk managers need to understand how a wide variety of risks might affect their own organizations. Emerging risk is an umbrella term for those new risks just on the horizon, very often on the edge of insurability but which have the potential to have a massive impact on organizations.

## A selection of emerging risks

We have selected a variety of emerging risks to consider in this chapter:

| | |
|---|---|
| Nature | Climate |
| | Earthquake |
| Technology | Cyber-risks |
| | GM technology |
| | Nanotechnology |
| Social and political | Terrorism |
| Infectious diseases | HIV/Aids, SARS |

These include existing risks such as climate, earthquake and terrorism, risks of which we have all been well aware for many years. Climate is included because we seem to be encountering more extreme weather patterns and worsening levels of catastrophic loss. For example, 2004 saw a massive $49bn in claims against property insurers – and this figure was due largely to significant weather-related losses.[3] Earthquake is on the list, because many areas of rapid urban growth are in earthquake prone areas. Unless stringent earthquake-resistant building codes are followed strictly significant economic loss and loss of life are likely to occur. In 2004 we also saw the Indian Ocean earthquake and subsequent tsunami which caused massive loss of life, significant property damage and substantial economic dislocation for many countries such as Indonesia, Malaysia, Thailand, the Andaman and Nicobar Islands, India, Sri Lanka, Bangladesh, the Maldives and even Somalia, Kenya and Tanzania. It now appears likely that an early warning system will be put in place to warn countries around the Indian Ocean of an impending tsunamis.

Terrorism, as we have mentioned, is not a new risk but its nature appears to have altered fundamentally since the 9/11 attacks. Since then we have seen the Madrid train bombings in March 2004 and then, of course, the bombing of three London underground tube trains and a London bus in a coordinated terrorist attack in July 2005. If it was needed, these provided a wake-up call about the ongoing threat of a terrorist atack and its ability to reach into the everyday lives of ordinary people, disrupt business and, in the case of Madrid, have far-reaching political implications.

Disease is not a new risk but HIV/Aids continues to be a massive human and economic catastrophe. The SARS outbreak in 2003 demonstrates how a new risk can seemingly appear from nowhere and attack many certainties almost overnight. Other risks have appeared only recently, for example, cyber risks and the risks emerging from GM technology and nanotechnology are only now being considered and understood.

## Definition

A review of risk management literature reveals that at present there is no universally accepted definition of emerging risk even though it is very much a current topic. According to *Reinsurance*, November 2003[4] emerging risk is about a 'seemingly endless list of unexpecteds'.

For our purposes it is necessary to break down the term into 'emerging' and 'risk'. There are various definitions of 'emerging' such as 'starting to exist' or 'the process of appearing'.[5] Others include 'to come forth from obscurity', 'to become evident' or 'to come into existence'.[6] 'Risk' can refer to 'the possibility of suffering harm or loss' and also 'a factor, thing, element, or course involving uncertain danger'.[7]

'Emerging risk' refers to new circumstances that have started to appear which have the possibility to produce unforeseen and probably negative outcomes. The definition extends to include existing circumstances that have taken on new or more significant manifestations.

While 'emerging risk' is real in the sense that we can identify the risk, in truth our view of potential consequences is clouded by a number of unknown variables. Also, of course, our fear of the unknown may mean that we exaggerate possible consequences and focus on the negative. This can be a significant problem. Enterprise risk management includes consideration of 'upside risk' and the danger is that if we focus on 'downside risk' new technologies are stifled and, consequently, new opportunities are missed.

## Identification and analysis of emerging risk

The beauty of traditional hazard risk is that identification, analysis and control are straightforward. The risks of a factory fire or car accident are easy to identify and the manifestation and the physical consequences of each are plain to see. Less obvious but still calculable are the financial consequences: the loss of revenue and profit arising from the fire, or the additional rental car costs while the car is being repaired. The probability of loss and potential severity under either scenario may be relatively easy to calculate. The risk manager of the company that owns the factory will be able to decide whether to retain, avoid or transfer the risk through the purchase of insurance. The property insurance underwriter will also have access to data to ascertain loss histories for the type of building, construction data, protections, the loss history and the management of the company. The motor fleet underwriter will be aware of loss histories for similar types of vehicle and average repair costs, as well as having access to good historic claims data for the company. In both examples, therefore, the risks are relatively straightforward to identify and analyse. Insurance premiums for each should also be relatively easy to calculate.

When we move across to emerging risk many of these certainties disappear.

### *Identification of emerging risk*

In fact, the very 'newness' of some emerging risks poses a problem in carrying out a proper identification of such risks. Returning to the example of the factory or the car we saw that identification of risk is relatively straightforward. Identifying the risks of climate change, for example, is much more difficult. It is becoming common to assume that we are going through some form of climate change involving global warming. Many people will produce anecdotes of warm weather in Europe in 2003 or extreme flooding in parts of Europe the previous year. Moving forward to a proper identification of the risk is much more complicated although statistics are available. Is the climate really changing for the long term or are we in the middle of a statistical 'blip'? What will happen in the nightmare scenario if the Gulf Stream shuts down? In any event whether the arguments in favour of global warming are accepted or not climate is constantly changing anyway. How will natural catastrophe modelling that has developed in the past 15 years respond to climate change? On the other hand there are also sceptics in the scientific community who contest whether global warming is taking place at all and point to significant gaps in the evidence on the subject.[8]

Other emerging risks contain even more unknowns. Nanotechnology involves 'the development of new materials and processes by manipulating molecular particles'[9] often a millionth of a millimetre in size. There has been some discussion that in the future nanotechnology will mean that we will have buildings or clothes that repair themselves although at this stage this seems many years in the future. At present it seems that too little is known about nanotechnology to carry out a full and proper identification and assessment of risks, although steps by industry, governments and other interested parties are now being taken to improve on this.[10] The danger of course remains that we end up discussing nightmare scenarios not based on reality and that proper consideration of risk is hijacked by scare stories about the world being taken over by a 'grey goo', or self-replicating robots. Identification of an emerging risk can therefore be extremely problematic, vague and sometimes emotive.

### Analysis of emerging risk

Analysis of the risks and potential losses involving our earlier examples of the factory or the car were relatively straightforward and we could make good use of historic data. These certainties disappear when we move on to the analysis of emerging risk. We may have identified an emerging risk but an analysis of likely manifestations of the risk and likely consequences is fraught with difficulty.

There may be interconnections and unintended consequences that we have not even considered. Systemic elements may well compound the difficulties in carrying out an analysis. An initial thought may be to avoid the risk but this ignores the fact that we often want to push out scientific boundaries. Also, to survive many organizations need continually to innovate.

Specific difficulties in analysing emerging risk will arise from:

1   *Lack of historic data.* In pricing insurance risks and in analysing risks in an organization great use is made of historic data. Even without historic data it is usually fairly easy to 'take a view' about an existing risk with some degree of accuracy perhaps using a risk matrix. In the absence of full historic data it is difficult to carry out this analysis on an emerging risk and it is also problematic to put together a risk matrix when there may be many unknown variables. Where we are considering naturally occurring events, then excellent historic data is available.

The problem that we will encounter is that the past may not be very helpful to a future, for example, where weather patterns become more unpredictable. For other emerging risks such as nanotechnology there will be no historic data from which to work.

2   *Modelling.* There have been huge leaps forward in modelling during the past 15 years. However, the accuracy of results from modelling is often only as good as the raw data used. In any event it is wise to use models judiciously and often an insurer or re-insurer will make use of several models to analyse the same risk. This problem will be compounded when we try to use modelling on an emerging risk where significant data are simply not available.

3   *New technology.* Where new technology is pushing out the boundaries of existing science, realistically how can all risks be modelled? There will often (possibly always) be some uncertainty.

4   *Human elements.* Certain emerging risks will contain a significant human element. In normal circumstances this is difficult to analyse. When we move across to emerging risk the problem will become compounded.

## Common features of emerging risk

Emerging risk contains several common features: the potential for catastrophic loss, interrelatedness and systemic felements, and also the potential for unintended consequences.

### *Catastrophic features*

If it were not for genuine concern that there are significant catastrophic features in emerging risk then our interest might be largely academic. Is this concern based on the reality of historic experience or purely on a perception that emerging risk will bring with it significant catastrophe? The answer to this is that it depends very much on the risk that we are considering.

In our discussion of emerging risk we have included the risks arising from natural events, specifically weather-related losses and earthquake. Both contain the potential for massive catastrophic loss. In terms of both one need look no further back than the year 2004. Weather-related events included Hurricanes Charley, Frances, Ivan and Jeanne in the US; and Typhoons Chaba, Songda and Tokage in Japan. In terms of earthquake, the Indian Ocean earthquake and subsequent tsunami was a catastrophe of massive proportions. The point in relation to both is that it seems that the potential impact of weather-related events and earthquake is worsening.

Turning to other risks, the events of 9/11 demonstrated beyond doubt that a terrorist attack could have catastrophic results. The Madrid train bombs showed also that a terrorist attack could have wide-ranging political implications.

Fortunately, the SARS outbreak in 2003 was contained. It nonetheless had the potential to be a major catastrophe. For other emerging risks it is difficult to judge. For example, are our fears surrounding GM technology and nanotechnology genuine? Have they been subject to too much hype by an overactive and imaginative media?

So is emerging risk real or illusory and just how significant is it? Anecdotally the concerns arising from emerging risk seem real. An investigation into recent historical experience of emerging risks reveals a very patchy picture. Some emerging risks (e.g. natural catastrophe) can produce real losses that can cause massive dislocation and economic cost (both insured and not insured). A review of other examples of emerging risk in relation, for example, to technology reveals that to a certain extent these are only just developing and it may be that our perception is perhaps far too coloured by the unknown and may even ultimately be damaging the development of the new technology itself.

## Interrelatedness and systemic elements

The very complexity of twenty-first-century life means that an incident in one place can have significant effects elsewhere. A recent example was the US power outage referred to above. A significant loss in a major city, such as a serious earthquake or large-scale terrorist attack could have major economic impacts elsewhere due to the interdependencies of the world economy. It has even been suggested that a significant earthquake in Tokyo could have the potential to trigger a worldwide recession.[11] The interconnectedness of modern life means that significant risks can have massive effects on the systems on which we rely. At the end of 1999 there was significant concern about the potential for harm that the Y2K issue might cause. The response to this systemic problem was a massive programme of IT upgrades. Fortunately, very few of the expected problems materialized. Our IT systems have frequently been plagued by viruses, e.g. the Mydoom virus which proved that a cyber-bug can have widespread effects and potential systemic risks are great. IT professionals are having to devise more and more sophisticated protections to guard systems against the potential impact of viruses. The fact that we live in such an interconnected world raises the prospect of systemic risk.

## Unintended consequences

There is a genuine concern in some quarters that because of difficulties in identifying and analysing the risks of new technologies we may also be opening ourselves up to unforeseen, irreversible or very expensive consequences. What, for example, might be the asbestosis equivalent for the next decade?

## Examples of emerging risk

### *Nature*

In a discussion about emerging risk it may seem rather strange to include discussion of the risks arising from natural events. After all such events have been taking place for billions of years. Why should we be unduly concerned about naturally occurring processes suddenly at the start of the twenty-first century?

We have seen that a significant problem in analysing emerging risk is that we have either few or no historic statistics on wish to base our judgements about the future. Often we have a smattering of historic data (if we are lucky), possibly well considered and researched studies about risk potentials. This is mixed in with a general perception that there may be a significant potential for something to go wrong in the future. If the worst does happen and something does go wrong then there is a genuine prospect for a major catastrophe. Subjective elements can therefore interfere with our perception of risk.

With the risks of natural events we have excellent historic statistics but it seems that these will not give us a full guide to the future with regard to weather-related events due to the impact of global warming. In this section we consider the risks of weather-related events with particular reference to climate change. We then look at the risks arising from earthquake.

### *Weather-related events: a growth in unpredictability*

A review of the recent past reveals that the frequency and severity of recent weather events seems to be worsening. In October 1987 the UK and France were struck by what became known as the Great Storm of 1987. In these two countries over 20 people were killed. Across southern England 15 million trees were lost. At the time this was viewed as the worst storm to hit the UK since 1703 with insured losses placed at almost $5bn (indexed to 2004).[12] Yet just over two years later in January 1990 the UK and France were struck by further storms where insured losses were higher at $6.6bn (indexed to 2004).[13] Autumn 2000 in the UK was the wettest since UK records began in 1766, with significant flooding in many areas. The year 2002 saw major floods in Germany, Austria and the Czech Republic. But in 2003 Western Europe was hit by an unprecedented heatwave which, it is estimated, killed 15,000 people.

During the past 20–30 years El Niño events are becoming 'more frequent, persistent and intense' compared with the previous 20–30 years.[14] In the recent past we have also seen major hurricane activity with Hurricane Hugo (1989), Hurricane Andrew (1992) and, of course, Hurricanes Charley, Frances, Ivan and Jeanne in 2004. In the same year Japan was also struck by Typhoons Chaba, Songda and Tokage.

We may be too close in time to say that the above reveal significant trends for the future. The cluster of four hurricanes hitting the US in 2004 was unusual but certainly not unprecedented. In 1916 and 1985, for example, six hurricanes made landfall on the US mainland.[15] Also, in respect of Florida, our view of weather-related catastrophes is going to be coloured by the fact that there has been a major influx of population during recent decades, massive property development and hence significant increases in insured values. In addition, any focus on catastrophes that hit the US, Japan and Western Europe will be affected by the fact that the propensity to insure and therefore insured values in these territories are among the highest in the world. Non-life premiums per capita in 2003, for example, in the US were $1,478.27, Japan $550.57 and the UK $1,040.66 compared with, for example, China $8.04 and India $2.88.[16] A feature of the 2004 Indian Ocean earthquake and tsunami was massive loss of life but, for the scale of catastrophe, a relatively low insurance impact. Estimated insured losses from this catastrophe are $4bn.[17] Non-life premiums per capita for 2003 were the following for some of the countries that were affected so tragically: Indonesia $6.58, Sri Lanka $7.35 and Thailand $25.73.[18]

Our continuing concern about weather-related losses is now being affected significantly by the expected impact of climate change during the next few decades. There is now extremely strong evidence that the global temperature is becoming warmer. It is estimated by the Intergovernmental Panel of Climate Change (IPCC) that during the twentieth century the Earth has warmed by about 0.6 degrees centigrade, with the rate of increase of the Northern Hemisphere surface temperature higher than during any other period in the last 1,000 years. The 1990s in the Northern Hemisphere was the warmest decade in the last 1,000 years. The IPCC predict that for the period 1900 to 2100 the global average surface temperature will increase by between 1.4 and 5.8 degrees centigrade and that the global mean sea level will rise by between 0.09 m and 0.88 m.[19]

Returning to the insurance impact it is interesting to note from the ABI report 'A Changing Climate for Insurance: a summary report for chief executives and policy makers'[20] that claims for storm and flood damage in the UK have doubled to over £6bn for the period 1998–2003, compared to the previous five years, and that this could triple by 2050.

Controversies about global warming continue but it seems likely that the climate changes detailed above have been caused by the emission of greenhouse gases (mainly carbon dioxide) into the atmosphere due to human activities. It is estimated that the pre-industrial atmospheric concentration of carbon dioxide of 280 ppm (parts per million) had increased to 368 ppm by the year 2000 and it is estimated that the concentration of greenhouse gases in the atmosphere will continue to increase during the twenty-first century.[21]

Worst-case scenarios talk of the shutting down of the Gulf Stream, which would have a major impact on the climate of Western Europe; or the West Antarctic ice sheet slipping off the continental shelf into the sea, which would have the potential to raise sea levels by five metres during the next few hundred years.

As an emerging risk the predicted changes in climate have the potential to make weather patterns more extreme and unpredictable. There are likely to be increased precipitation with increased risks of flood in some areas of the world and an increased risk of drought in others.

It seems that the unpredictability that has crept into weather-related losses during the last 15–20 years is set to continue and may become more pronounced.

### Earthquake

Insurers and risk managers often focus on the impact to physical property or economic losses arising from major loss events. However, 2004 showed how a major earthquake can have a massive impact on human life. Of the ten worst catastrophes during the period 1970–2004 in terms of fatalities it is notable that seven out of ten were caused by earthquake.[22] The true number of people who lost their lives in the December 2004 Indian Ocean earthquake and tsunami will never be known but estimates of dead and missing are well in excess of 200,000.

Again, earthquake is not a new risk so why are we treating it as an emerging risk? Much of the world's current population growth is taking place in urban areas. For example, it is estimated that 40 out of 50 of the fastest growing cities in the world are in earthquake-prone areas. In addition it is believed that in 50 years time one-third of the world's population will live in volcanically or seismically active areas.[23] The very fact that such large numbers of people are living in such vulnerable areas increases the risk of significant loss of life dramatically.

### Technology

Often, consideration of new technology from the risk and insurance perspective will begin with either little or no knowledge. This may lead the risk manager to seek to avoid the risks of the new technology and for the insurer to issue a blanket exclusion. As the state of knowledge develops, usually incrementally, and as risk professionals engage in the debate surrounding the new technology they will seek to identify whether it really presents a new set of risks or an old set of risks dressed differently. Finally, when a new technology becomes part of the established order then insurance may become routinely available. Cyber-risks, GM technology and nanotechnology represent three stages in the process. Organizations have strong risk management procedures in place to deal with their information technology risks and cyber-insurance is now relatively easily available. With GM technology insurers are grappling with the issues without having reached a firm conclusion on how to move forward. The risk management aspects of nanotechnology have only very recently started to be considered although there does seem to be a willingness by many parties to grasp these issues at an early stage. This will be to everyone's advantage.

*Cyber-risks*

There are two main reasons from the emerging risk perspective that we are interested in cyber risks:

1   the critical importance of the cyber-economy – modern economies simply cannot function without computer networks, software etc.;
2   the cyber-economy hits the very issue of systemic risk and the possibility for an incident in one area having massive impact elsewhere.

The interconnectedness of computer networks and the widespread use of the internet mean that there is great vulnerability to computer viruses and coordinated attacks by hackers. Good statistics on general cyber-losses are not available and it is likely that current cyber-losses are dwarfed by losses that arise from natural catastrophes.

Yet, to a large extent, cyber-losses reach the very hub of our concerns about emerging risk in that they reflect the interconnectedness of modern life and the potential for massive systemic loss.

*GM technology*

As an emerging risk the debate about GM crops means that this is one of the most emotive subjects to consider. The debate ranges from the academic press to the tabloids with stories about 'Frankenstein foods' keeping the subject high in the public consciousness and with various interest and pressure groups and, of course, companies taking widely divergent views. In addition it is interesting that various countries take markedly different approaches – the US and EU, for example, take widely diverging approaches. It is necessary for insurers and risk professionals to stay away from the emotional aspects of the debate in order to gain a greater understanding of GM technology.

On the one hand, the debate centres about whether or not GM foods are actually necessary. It is suggested that GM technology is not needed to meet current UN targets of halving hunger by 2015. By 2050 the situation could be significantly different, with food production having to increase by 60 per cent to meet population growth projections.[24]

Based on this it would seem that there will be a greater imperative to adopt GM technology and while it may not be critical for Europe where many countries are likely to face static or declining populations in the next decades, in developing countries there will be added pressure to adopt GM technology.

According to Jacques Diouf, Director General of the UN Food and Agriculture Organisation:

> Such a situation will require intensified cultivation, higher yields and greater productivity. We will have to use the scientific tools of molecular biology, in particular the identification of molecular markers, genetic mapping and gene transfer for more effective plant enhancement, going beyond phenotype-based methods.[25]

But, on the other hand, it has been suggested that some insurers categorize GM crop risks alongside that of the thalidomide drug, asbestos and terrorism.[26] On the surface these may seem extreme views but it should be borne in mind that there are still many unknowns about the risks of GM technology. Two fundamental issues for insurers are that, first, generally it is thought that not enough is known about the long-term risks of GM technology and, second, the vagueness surrounding legal liabilities arising from the use of GM technology means that insurers may be reluctant to be fully involved until there is more certainty.

The risks of GM technology have been identified as risks to health, the environment and biodiversity. The lack of conclusive research into the risks of GM technology means that the issue of insurability has yet to be resolved although as its use becomes more widespread there will be a strong imperative for the insurance industry to grapple with this issue and reach a firm conclusion.

### Nanotechnology

Nanotechnology has only recently appeared on the radar screens of insurance and risk professionals. Even now the debate on the impact of nanotechnology is at its earliest stages and several years behind consideration of the risks arising from GM technology. The challenge will be to ensure that the debate surrounding the risks of nanotechnology avoids becoming emotive so that these can be given proper and full consideration.

'Nano' is believed to be derived from the Greek word for dwarf and this gives a good idea of what this new technology involves. According to The Royal Society and The Royal Academy of Engineering, nano-technologies 'are the design, characterisation, production and application of structures, devices and systems by controlling shape and size at nano-metre scale'.[27] One nanometre (nm) equals one-billionth of a metre or approximately the length of ten hydrogen atoms. A human hair, on the other hand, is 80,000 nm wide.[28] Thus, when considering particles at the nanoscale we almost have to suspend disbelief.

A few years ago nanotechnology would have been very much in the realm of science fiction and was largely unknown outside a small circle of scientists and science journals. Now it is rapidly turning into a multi-billion dollar industry. It is highly likely that during the next few decades nanotechnology will affect most aspects of our lives including the food we eat and the clothes we wear.

The growing significance of nanotechnology can be seen from the fact that it is estimated that in 2004 nanotechnology R&D by govern-ments, companies and venture capitalists was in excess of $8.6bn.[29] In 2004 the US government spent $1.6bn on nanotechnology and it has been described as 'the largest federally funded science initiative since the country decided to put a man on the moon'.[30] Nanotechnology represents a massive leap forward in manufacturing techniques and could represent a new industrial revolution. Opportunities arising from nanotechnology include lighter and stronger materials with benefits in many industries, such as communication and information technologies and others such as pharmaceuticals.

A feature of nanotechnology is that at such tiny sizes particles are no longer subject to the laws of classical physics but become subject to the laws of quantum mechanics and can therefore demonstrate different capa-bilities. This is where the concerns about the risks of nanotechnology arise. Where a new manufacturing technique is being used in which particles can behave differently than would normally be expected are we intro-ducing something that is too unpredictable and that could be dangerous to human health and the environment notwithstanding its significant bene-fits? This is some way off the world being taken over by self-replicating robots but we still need to understand fully how, for example, the body is likely to react to inhalation or ingestion of nanoparticles as well as likely effects on the environment. An added complication is that the smaller a particle becomes the more reactive the substance that the particles consti-tute is likely to become. There has been some speculation that nanotech-nology may represent as much of a threat as, say, asbestos. This may be too gloomy a scenario but it remains the case that more needs to be under-stood about the risks of nanotechnology notwithstanding the massive benefits these new manufacturing techniques will bring.

### Social and political

#### Terrorism

At first sight it may seem strange to include terrorism in a discussion about emerging risk. After all, terrorism is not a new phenomenon: the

UK has lived with the terrorist threat for many years. However, from the late 1980s there was an increase in the severity of terrorist attacks, which also started to become much more daring, such as the Lockerbie bombing in 1988 in which a Pan American Boeing 747 was blown up killing 270 in the air and on the ground, and bombings in London in April 1992 and April 1993, Oklahoma in April 1995, Manchester in June 1996, Columbo International Airport and Sri Lanka in July 2001. The infamous 9/11 attacks in the US where four commercial airliners were hijacked and where two were crashed into the World Trade Center, one into the Pentagon and one of which crashed in Pennsylvania, and where over 3,000 people were killed, stands out as by far the most significant.

The justification for including terrorism as an emerging risk is that it is an old risk that has recently taken on significantly new dimensions (particularly since 9/11) incorporating both unpredictability and the possibility of significant property loss and loss of life.

The issue of public/private partnerships as an insurance response to terrorism risks is not new. Pool Re, where the UK Government acts as Pool Re's re-insurer of last resort, has been in operation in the UK since 1993. But since 9/11 there has been an expansion of public/private partnerships to other territories, for example the Terrorism Risk Insurance Act 2002 (TRIA) in the US; in France, Gareat (Gestion de l'Assurance et de la Réassurance des Risques Attentats et Actes de Terrorisme) – a government-sponsored pool; and in Germany, Extremus Versicherungs AG. There have also been further developments to Pool Re such as the expansion of cover to an all risks basis, inclusion of cover for nuclear contamination and that premiums for primary insurance cover can be set by insurers themselves rather than set by Pool Re.

The 9/11 attacks also provided a wake-up call to business and caused many people either to review the adequacy of existing business continuity plans or to implement new plans encompassing emergency response, crisis management and business recovery. This was particularly noticeable for companies with critical operations in city centre locations.

In the future terrorism is likely to remain high on the agenda for most risk managers. This was given added impetus by the bombings of crowded commuter trains in Madrid in March 2004 and then of tube trains and a bus in London in July 2005. Steps are now being taken to model terrorism risks.[31]

### Infectious diseases

As emerging risks, infectious diseases reflect potential catastrophes at both the social and economic level. HIV and Aids are truly devastating to many countries. The SARS outbreak in 2003 demonstrates how an infectious disease can, in a relatively short space of time, appear and cause significant loss of life and massive disruption.

*HIV/Aids*

HIV and Aids came to the public's attention around the mid-1980s. In 2004 it was estimated that 39.4 million people were infected with HIV, up from 36.6 million in 2002; 3.1 million people died from Aids in 2004 and 4.9 million people acquired the HIV virus.[32] HIV continues to be a massive issue particularly for countries of Sub-Saharan Africa where it is estimated that 25.4 million people live with HIV (or 60 per cent of the world total).[33] The economic, social and, also, personal consequences for the populations of certain countries are truly catastrophic.

In the UK, HIV and Aids continue to be a major problem. In 2002 there were 5,711 HIV diagnoses and cumulative diagnoses since testing began (in 1982) now stand at 59,497.[34] Since 1982, 12,760 people are reported as having died from Aids.[35] The number of deaths from HIV-associated diseases has however reduced significantly from 1,516 in 1995 to 235 in 2001.[36] The number of people with diagnosed HIV in the UK is set to increase by 47 per cent between 2000 and 2005.[37] Thus the increase in diagnoses and reduction in the number of deaths from HIV suggest that the number of people living with the disease and receiving the necessary retro-viral treatments is likely to continue.

The insurance industry's response to HIV is evolving to reflect the fact that current treatments can, to a certain extent, manage the condition and help prolong life and delay the onset of Aids.

*SARS*

The SARS (Severe Acute Respiratory Syndrome) outbreak in 2003 represents an emerging risk that has only recently appeared. The symptoms of SARS include high fever and dry cough together with shortness of breath or breathing difficulties. Crucially for the rapid spread of SARS, the incubation period is just 3–6 days.

The first case of SARS was reported in Vietnam in February 2003. In truth however it seems that the SARS infection had appeared in China several months earlier. The virus spread quickly and cases were soon reported in China, HK, Taiwan, Singapore, Canada and the US.

When SARS first appeared it was quite rightly viewed as a major risk to world health. Vigorous efforts by several governments, particularly the Chinese and Canadian, prevented a much wider epidemic. A significant feature in the rapid spread of SARS around the world, i.e. from China to Canada and the US, arose because of the prevalence of air transport. It has been speculated that if the SARS outbreak had taken place in a poor/developing country, it could turn out like the influenza pandemic of 1918/19 as a result of which it is estimated that half a billion people were infected and that 30 million people died – more people than were killed in the First World War.[38] For a short while it seemed that SARS had the potential to be a major catastrophe. But there still remain ongoing concerns that an influenza pandemic, perhaps for example a strain of avian influenza, may take place at some stage in the near future and that its rapid spread globally could mean that the response capabilities of healthcare systems become compromised.

## Insurance/risk transfer

It is crucial for the insurance industry to gain an understanding of emerging risk. Insurers price for known risks. When they are caught unawares they can pay out huge losses for risks that either haven't been priced for at all or haven't been priced for adequately. We have seen this with the 9/11 terrorist attacks in the US. The scale of the losses was so unforeseen that insurers and re-insurers faced massive hits to their balance sheets. Before that we have seen the impact on the insurance industry of environmental issues and we are still seeing the continuing impact to the industry of asbestosis. Risks that were either not considered at all or barely considered resulted in a massive impact on the insurance industry. In the future a major systemic loss could severely impact the insurance industry.

When understood and priced for correctly the catastrophic elements of emerging risk could be an opportunity for both insurers and re-insurers and also the capital markets with ART products.

### Insurance response to emerging risk

Because emerging risk refers to myriad different risks, some of which have only recently appeared, the response of the insurance industry has varied depending on the risk. Emerging risk by its very nature operates at the edge of insurability. Lahnstein, in 'The Insurability of New Liability Risks', points out the usual criteria for insurability of private insurance risk as follows: the risk has to be random, unambiguous, estimable, independent and of appropriate size.[39] For certain emerging risks meeting these criteria will be at best problematic and at worst impossible. Often a particular risk will have been excluded from policies once the risk became apparent but over time as knowledge of these risks has improved and some of an underwriter's natural scepticism has been replaced with better information then this approach is relaxed. It may be possible, for example, after time to 'buy back' the cover or it may be that specific policies have been developed.

As we have seen, earthquakes, hurricanes, floods and other natural catastrophes are not new but new features are that the scale and frequency of such occurrences are increasing. Vulnerabilities are also increasing with significant population and value increases in areas susceptible to catastrophic loss. The insurance and re-insurance industry markets clearly continue to offer covers for earthquake, hurricane and flood but this has been coupled with a desire and need to understand these risks as evidenced by the growth of the catastrophe modelling industry during the past 15 years. At the same time other risk transfer arrangements have developed to manage natural catastrophes, e.g. public/private partnerships in certain countries and pooling arrangements to supplement conventional insurance and re-insurance capacity. Examples would be the Florida Hurricane Catastrophe Fund or the Turkish Catastrophe Insurance Pool.

Cyber-risks are usually excluded from traditional property/casualty insurance covers but demand for specific cyber-risk policies, particularly in the US, continues to grow apace. This is also now becoming apparent in other territories outside the US, particularly the UK. While demand was initially focused on third-party risks such as copyright infringement, defamation, invasion of privacy and errors and omissions, this has now been expanded to first-party covers such as electronic theft of money, securities, loss of or damage to the insured's network and business interruption. It may be that in 10–15 years time a company will, as a matter of course, buy a cyber-risks policy alongside its current portfolio of property/casualty covers. As with IP insurance covers where first-party products are being developed to meet new exposures there is a growing realization that first-party issues can also be massively significant.

Other technological risks as far as the insurance industry is concerned are very much in their infancy and the insurance industry has yet to formulate its response; and it is interesting to speculate where the incremental developments may lead during the next 10–15 years. With world population expected to increase by 50 per cent by the year 2050 there will be a significant imperative to use GM crops and therefore pressure for the insurance market to develop its response and offer suitable risk transfer products.

There have been some recent developments with regard to infectious diseases such as HIV/Aids. A very limited number of insurers, for example, have introduced life insurance products aimed specifically at people with HIV. Also in 2004 the Association of British Insurers issued its 'Statement of Best Practice on HIV and Insurance' aimed at assisting insurers to assess applications where HIV may be an issue and also to avoid intrusive and inappropriate questions.[40] Thus, the insurance industry's response to the issue of HIV and Aids continues to develop and is becoming more constructive.

So except for risks that have been prevalent for some time the insurance response has usually been to exclude but then modify the approach by removing this exclusion incrementally or developing new products when the risk is understood better. This may coincide with a technology obtaining mainstream acceptance such that it becomes routine. Emerging risk clearly represents both a significant threat to the insurance industry particularly but also a significant opportunity in the form of new product opportunities. Also, there are better opportunities to understand some emerging risks through more sophisticated modelling techniques.

## Difficulty in quantifying risk

A fundamental problem for the insurance industry in its approach to emerging risk may be the lack of any significant quantitative data and other historic information in reaching pricing decisions for new risk. If this is combined with strong suspicion that there is potential for significant loss this may make it difficult to build a good business case for insuring emerging risk, particularly in view of more and more stringent requirements from capital providers, rating agencies and regulators. It is likely that the insurance industry will take an incremental approach to insuring completely new risks, for example, arising from new technologies.

### *Increased use of modelling techniques in the insurance industry*

The increase in natural catastrophes has led to a rapid growth in the past 15 years of an interest in modelling techniques by insurers and re-insurers. Modelling is becoming increasingly expansive and has now even extended to terrorism.

The growth in the use of modelling can be specifically attributed to two reasons. The increased use of ART to 'insure' catastrophic exposures. Rating agencies and capital providers have required a far more scientific approach to risk measurement than previously and usually require these catastrophic risks to be modelled. In addition, the impact of Hurricane Andrew in 1992 demonstrated that the use of historic data to quantify exposures could have severe flaws. In the example of Hurricane Andrew it seems that insurers pricing decisions at the time were based on hurricanes experienced during the previous 30-year period. This period was, in fact, unusual and losses were significantly lower than the long-term trend. The result was that the losses sustained in Hurricane Andrew were often inadequately priced with the result that certain insurers suffered massive losses. It would seem that the cluster event of four hurricanes making landfall in the US in 2004 will give the modelling industry a large amount of data to further refine its models particularly with regard to such cluster events.

Typically there are three elements to catastrophe modelling consisting of a hazard, damage and loss module. Modelling is now becoming far more sophisticated and now complex meteorological forces are being modelled. As mentioned, terrorism, which includes the major unknown variable – the human element, is starting to be modelled.

A feature of catastrophic events in the last few years is the speed with which modelling companies can produce reasonably accurate estimates of aggregate loss. It is too early to speculate whether modelling will expand to other areas of emerging risk but if the risks arising from new technology could be modelled in any meaningful sense this would significantly assist risk managers and insurers in considering these risks.

### Business interruption and business continuity planning

Business interruption frequently occupies one of the top slots in surveys of risk managers of the major risks they are most concerned about. Business interruption losses can often dwarf property damage losses.

Business interruption insurance covers the loss of gross profit following an insured loss. The policy will also normally cover extra expenses incurred following the loss to continue trading. There may also be extensions such as customers' or suppliers' extensions to cover, for example, dependencies which are often critical.

In the recent past the insurance industry has suffered heavy business interruption losses. In the context of emerging risk, business interruption can be a prominent feature and figured heavily, for example, in losses following 9/11 through widespread disruption to business, and following the SARS outbreak because of cancellation of events and denial of access.

The 9/11 attacks gave a significant boost to the development of business continuity planning when policyholders realized that while business interruption is important, it is critical to become operational as soon as possible following a significant loss. It seems likely that business continuity planning will be massively important as a response to emerging risk.

### Liability insurance

The vast amount of unknowns surrounding emerging risk means that this is a major issue for liability insurers. Seemingly innocuous issues today can often return to bite liability insurers tomorrow and there are numerous examples where this has happened. Thus, liability insurers have previously unwittingly incurred massive claims for activities that they had not contemplated or were either unaware or not fully aware of the risks that they were underwriting. Examples might include asbestos, which was initially considered a new wonder building material. Also, insurers have inadvertently had to deal with significant pollution claims when, at the time of writing such policies, environmental risks were not considered.

Liability insurance encompasses a broad range of policies that insurers offer to protect against legal liability claims for such matters as property damage or bodily injury. They include policies that cover public liability, products liability, professional indemnity, errors and omissions, directors' and officers' liability, employer's liability and workers' compensation. A feature of such policies is that often (particularly where written on an occurrence basis) insurers may end up having to deal with claims made a long way in the future, i.e. such policies often contain a significant 'tail'.

As far as the liability insurer is concerned matters to be considered when dealing with a potential emerging risk might be:

- The fact that there are significant unknowns, e.g. are there potentials for physical damage and bodily injury? We mentioned earlier the possible risks of inhalation or ingestion of nanoparticles.
- Lack of historic data.
- Increasing tendencies to make a claim against another party when something goes wrong.
- The changing legal environment and resultant uncertainty: in the first place the law is constantly expanding incrementally due to the impact of case law; a further issue is that the legislative response to a new technology is often to impose strict liability on producers for bodily injury irrespective of fault; the danger is to inadvertently underwrite all the risks of a new industry irrespective of fault.

It is difficult to predict how insurers will respond to insuring the legal liabilities arising from new technologies such as biotechnology and nanotechnology. It may be that cover for such risks is initially restricted or excluded entirely but that specific bespoke covers aimed at such technologies are developed, as happened with cyber-risk policies. In any event, a review of insurance and risk management literature reveals that the insurance industry is more than willing to enter the debate on how risks arising from new technologies can be transferred.

### Alternative risk transfer (ART)

Alternative risk transfer or ART refers to alternative, non-traditional transfer of (re)insurance risks usually placed with the capital markets and can include catastrophe bonds, securities, insurance derivatives and contingent capital. It is worthwhile to consider the extent to which ART solutions will either complement or take the place of traditional (re)insurance solutions to emerging risk.

Our consideration of emerging risk has demonstrated that it can be a difficult and slippery fish to catch in the sense that identification is problematic; once identified there are difficulties in analysing the significance of the risk and then managing such risks can be difficult. Where the emerging risk is a well-known risk that has taken on new dimensions, e.g. natural catastrophes or terrorism, then issues can be more clear-cut in that quantification, analysis and modelling can usually take place. Where the emerging risk involves a new technology then this approach can be far more difficult.

We have seen that 'emerging risk' is also a 'catch all' term that refers to a wide variety of risks such as natural events, social and political, technological and risks of infectious diseases.

A brief consideration of ART since these solutions started to be offered in the mid-1990s reveals a steady growth in the use of such products, particularly in the context of natural catastrophes and often as a complement to, and sometimes a substitute for, traditional re-insurance approaches for the transfer of risk. While there has been steady growth it should be realized that the total of ART transactions is dwarfed by the value of standard insurance and re-insurance transactions. It was thought, for example, that when the traditional insurance and re-insurance markets hardened during 2000–2002, ART would become a credible alternative to property/casualty insurance. Market analysts, however, estimate that during 2002 just $2bn of new catastrophe bonds were issued while the amount of equity capital that went into the world's property and casualty insurers was a multiple many times greater.[41] It may be that the complexity of such transactions or simply the lack of familiarity with such products has meant that re-insurance buyers have chosen a more traditional approach. It may also be that we have not yet seen the full potential for ART transactions in the future as they continue to grow and offer innovative risk transfer solutions.

ART is likely to continue to grow and in the context of emerging risk is likely to play a role alongside the traditional market in transferring both natural and man-made catastrophes.

## Conclusion

It is difficult to reach any firm conclusions on emerging risk save to say that the risk universe is becoming increasingly complex. Insurers and risk managers must continue to monitor emerging risk and simply take the time to learn about new risks and debate the issues surrounding identification, likely manifestations and likely consequences, bearing in mind interdependencies, systemic elements and unintended consequences.

The challenge will be for insurers and risk managers to stay at the leading edge of this debate on emerging risk. The easiest thing in the world is for the insurance industry to deal with emerging risk by introducing a wider range of exclusions to their policy covers. The challenge for the insurance industry will be to understand each emerging risk and develop a strategy on how the industry can respond to each new emerging risk as it appears. At the same time and to avoid a repeat of past mistakes the industry will need to be cognisant of the requirements of shareholders, capital providers, regulators as well as customers. The tools to recognize these are improving, for example, through improved risk management methodologies such as the AIRMIC Risk Management Standard[42] and continued development in the sophistication of catastrophe modelling.

Emerging risk represents a serious threat for insurers and risk managers to consider and, at the same time, if handled correctly, could also represent a great opportunity.

## Notes

1 'Emerging risks: a challenge for liability underwriters', Jurg Spuhler, Swiss Re (2003).
2 'Emergent risks', Research Paper by Bryan Richardson and Peter Gerzon, Institute of Risk Management (2005).
3 'Natural catastrophes and man-made disasters in 2004', Swiss Re, sigma No. 1/2005.
4 'Preparing for the worst', Marc Jones, *Reinsurance* (November 2003).
5 See Cambridge Dictionaries Online, www.dictionary.cambridge.org.
6 See Dictionary.com – ultimate source, *The American Heritage Dictionary of the English Language*, 4th edn, Houghton Mifflin Company (2000).
7 See Dictionary.com – ultimate source, *The American Heritage Dictionary of the English Language*, 4th edn, Houghton Mifflin Company (2000).
8 See, for example, 'Climate change: menace or myth?' by Fred Pearce, *New Scientist* (12 February 2005).
9 'Nanotechnology: tiny technology that may create problems', Andrew Bolger, *Financial Times* (1 June 2004).
10 See, for example, 'Nanotechnology: small matter, many unknowns', by Annabelle Hett, Swiss Re (2004); and also 'Nanoscience and nanotechnologies', *The Royal Society & The Royal Academy of Engineering* (July 2004).
11 'Megacities – megarisks – trends and challenges for insurance and risk management', Munich Re (2004).
12 'Natural catastrophes and man-made disasters in 2004', Swiss Re, sigma No. 1/2005, Table 9, p. 34.
13 'Natural catastrophes and man-made disasters in 2004', Swiss Re, sigma No. 1/2005, Table 9, p. 34.
14 'Climate change 2001: synthesis report – summary for policymakers', Intergovernmental Panel on Climate Change (IPCC) (2001).
15 See NOAA Technical Memorandum NWS TPC-1 'The deadliest, costliest, and most intense United States hurricanes from 1900 to 2000', by Jerry D. Jarrell (retired), Max Mayfield and Edward Rappaport, NOAA/NWS/Tropical Prediction Center, Miami, Florida; and Christopher W. Landsea, NOAA/AOML/Hurricane Research Division, Miami, Florida.
16 Figures obtained from *Axco Global Statistics*, Axco Insurance Information Services Ltd.
17 See Willis Re Panorama – 'Catastrophes >> in the news #4: as @ 18 February 2005', Willis, 2005.
18 Figures obtained from *Axco Global Statistics*, Axco Insurance Information Services Ltd.
19 'Climate change 2001: synthesis report – summary for policymakers', Intergovernmental Panel on Climate Change (IPCC) (2001).
20 'A changing climate for insurance', Dr Andrew Dlugolecki, Association of British Insurers (2004).
21 'Climate change 2001: synthesis report – summary for policymakers', Intergovernmental Panel on Climate Change (IPCC) (2001).
22 'Natural catastrophes and man-made disasters in 2004', Swiss Re, sigma No. 1/2005, Table 10, p. 35.

23   See 'Emerging systemic risks in the 21st century: an agenda for action', OECD, referring to Randall, J.B., D.L. Turcotte and W. Klein eds, 'Reduction & predictability of natural disasters', Volume XXV of *Santa Fe Institute Studies in the Science of Complexity*, Addison Wesley (1996).

24   'GM food "essential" to meet future needs', John Mason, www.ft.com (15 June 2004).

25   Quoted in 'GM food "essential" to meet future needs', John Mason, www.ft.com (15 June 2004).

26   'GM crops "uninsurable"', Symon Ross, *Insurance Day* (8 October 2003).

27   See 'Nanoscience and nanotechnologies' (Chapter 2), *The Royal Society & The Royal Academy of Engineering* (July 2004).

28   See 'Small wonders: a survey of nanotechnology', *The Economist*, (1 January 2005).

29   See 'Small wonders: a survey of nanotechnology', *The Economist*, (1 January 2005) (referring to estimate by Lux Research).

30   See 'Small wonders: a survey of nanotechnology', *The Economist*, (1 January 2005).

31   See Risk Management Solutions Press Release, 'RMS responds to heightened international terrorism threat with launch of first global terrorism model' (23 September 2004); and also Gordon Woo, 'Quantifying insurance terrorism risk', Chapter 14 in Morton Lane ed. *Alternative Risk Strategies*, Risk Waters Group Ltd (2002).

32   See 'Aids epidemic update: December 2004', United Nations Programme on HIV/Aids and World Health Organisation, (2004).

33   See 'Aids epidemic update: December 2004', United Nations Programme on HIV/Aids and World Health Organisation, (2004).

34   See 'Numbers of people in the UK with HIV', Terrence Higgins Trust, www.tht.org.uk.

35   See 'Numbers of people in the UK with HIV', Terrence Higgins Trust, www.tht.org.uk.

36   See 'Numbers of people in the UK with HIV', Terrence Higgins Trust, www.tht.org.uk.

37   See 'Numbers of people in the UK with HIV', Terrence Higgins Trust, www.tht.org.uk.

38   See 'The truth about SARS infection', www.globalchange.com.

39   'The insurability of new liability risks', Christian Lahnstein, *The Geneva Papers on Risk and Insurance* (Vol. 29, No. 3, July 2004).

40   See 'Statement of best practice on HIV and insurance', Association of British Insurers, October 2004.

41   See Chapter 1 'Reinsurance versus other risk-transfer instruments: the reinsurer's perspective', in Morton Lane ed. *Alternative Risk Strategies*, Risk Waters Group Ltd (2002).

42   See 'Risk management standard', AIRMIC, ALARM, IRM, 2002.

# 3 Intangible assets, risk management and insurance

## Bringing all minds together

*Matthew R. Hogg*

### Innovation and intangible assets

Innovation is nothing new. Humankind has been innovative in one way or another since time immemorial; it is a state of introducing something new but not necessarily introducing that concept to anyone else. In today's language, however, innovation has taken on further meaning in the common parlance of our commercial world. Innovation has become much more than 'creation'; it has become synonymous with successful exploitation of new ideas and this success is measured in financial terms. Innovation for financial return lies in the incorporation of new technology, business methods and best practice. As the UK's Department of Trade and Industry states, such innovation requires companies 'to bring to the market a stream of new and improved, added value, products and services that enable the business to achieve higher margins and thus profits to re-invest in the business'.[1] Such innovation usually emanates from a considerable investment of human and financial resources: Research and Development (R&D).

Global R&D is big business. It is calculated that the top 700 international R&D spending companies pump £204.6 billion into their R&D departments in a single year. An estimated 38.3 per cent of this expenditure, not unexpectedly, is to be found in the US, with Europe a little behind contributing 35.9 per cent. Astonishingly, Japan, continuing to make efforts to innovate its way clear of economic inertia, holds a 22.4 per cent slice of the total top 700 R&D spend. The significance of such expenditure can be seen when one considers that R&D effectively represents 4.9 per cent of sales in the US, 4.2 per cent in Japan, and 3.7 per cent in Europe. Sadly, the UK's R&D investment sits somewhere around the 2.3 per cent mark, an acknowledged problem in government and industry, with responding initiatives now emerging to support innovation in the UK.[2] The companies obtaining the most patents for inventions also show a clear US and Japanese bias (see Table 3.1).

The fruits of innovation are more and more clearly associated and accepted as producing a diversity of intangible assets for the organization in question. The origin of intangible assets, as a term falling within financial terminology, may not be quite as old as innovation, but has certainly not been given much respect and attention until the last 25 years or so. In fact shareholders, financial institutions, investors, accountants, board members, insurers and risk managers are arguably still not giving intangible assets the full respect they deserve. Times are changing.

An intangible asset is often considered as an identifiable, fixed, non-monetary asset without physical substance that is held under the control of an organization whether through natural or legal rights. Typically, intangible assets are considered to fall into two not entirely distinct groups; goodwill and IP. Goodwill has historically been simply a balance sheet term to help explain the amount by which a purchase price of a company would exceed the net tangible assets of that acquired company. Goodwill in its broader sense, however, incorporates and recognizes the economic value of a company's internally developed, as well as purchased, intangible assets and includes its management and intellectual capital, R&D, customer-based loyalty, competitive position, and also indicates the potential for future success and the brand's strength. A good example of this in practice is to consider that the S&P 500 currently trades at around 300 per cent of its book value. Investors do not believe that the balance sheets of such companies reflect their real economic worth. IBM, for example, may have net assets of around $28 billion, but this is a drop in the ocean compared to an equity market value of somewhere around $155 billion. Similarly Microsoft's market value is somewhere around 20 times its book value. PriceWaterhouseCoopers estimates that around 74 per cent of the average purchase price of acquired companies in 2003 is attributable to intangible assets and goodwill.[3]

*Table 3.1* Top ten ranking patent recipients

|  | US patents | | UK patents | |
|---|---|---|---|---|
|  | *Company* | *Domicile* | *Company* | *Domicile* |
| 1 | IBM | US | NEC | Japan |
| 2 | Matsushita Electric Industrial | Japan | Hewlett-Packard | US |
| 3 | Canon Kabushiki Kaisha | Japan | Samsung Electronics | Japan |
| 4 | Hewlett-Packard | US | Schlumberger | US/France |
| 5 | Micron Technology | US | IBM | US |
| 6 | Samsung Electronics | South Korea | Baker Hughes | US |
| 7 | Intel Corporation | US | Ericsson | Sweden |
| 8 | Hitachi | Japan | Motorola | US |
| 9 | Toshiba Corporation | Japan | Visteon | US |
| 10 | Sony Corporation | Japan | Ford | US |

Source: Publicly released data from the US Patent and Trademark Office and the UK Patent Office. The Japanese Patent Office does not release such information.

There has been a significant shift in the value and dependency upon intangible assets over the last 25 years with intangible assets estimated to amount at most to only 20 per cent of market value in the late 1970s. This is a natural result of our 'knowledge economy', with technology, global expansion and service improvements allowing companies to do more with fewer resources. It is also due to the exponential growth of brands as a marketable differentiator to consumers. It is estimated that the value of the top three brands in the world, Coca-Cola, Microsoft and IBM, range from over $50–70 billion, and their fickle nature can mean that brands can show huge growth or loss, estimated examples being Apple with an estimated 24 per cent increase between 2003 and 2004, or Nokia down 18 per cent in the same period.[4] R&D expenditure is, of course, often the currency for such improvements.

IP is one form of intangible assets but should really be considered separately from goodwill. Goodwill is the 'softest' form of intangible asset, in that it incorporates the whole gamut of concepts that attempt to explain that the value of an organization extends well beyond its book value. IP rights have been drawn from this mire in recent years because of their clearer association to property rights as defined by the relevant localized national IP law. Although clearly still ethereal in that they can often not be seen or touched, there has been a common aim to set parameters for IP creation and classification lending itself to penning details of the IP right on paper as much as possible. This concept of IP rights, although not new (the first known patent, for example, was granted in 1421 in Florence, Italy, with England following soon after in 1449), has been refined or extended by many countries in later years. Being a property right, IP is often bought, sold, rented or hired. Global accounting developments in recent times have also considered IP rights distinct from goodwill for amortization purposes.

There are a number of IP rights that can be created in an organization going about its daily activities and these will be particularly apparent for a highly innovative company. The products and processes of an organization may arise from trade secrets and know-how; the Coca-Cola formula is protected as a trade secret, but it may be more practical to apply for a patent or utility model. A patent system has been described as 'the fuel of interest to the fire of genius'[5] because of the limited monopoly a patent grants to the inventor to prevent another from making, using or selling an invention. Patents protect innovation that has a technical element, but by gaining a patent the inventor shares this technology with the general public to further society's understanding. A patent right typically lasts for 20 years, but is dependent upon the legal system.

Distinctive signs, brands and, in the public's eyes, names are protected through the grant of a trademark. A trademark is effectively a badge of origin that distinguishes one thing from another and therefore to gain the property right it must be distinctive in that area of goods or services. A trademark can often last as long as the owner wishes to maintain that mark, which often also means paying regular fees to the relevant governmental office if it is a registered trademark.

Literary and artistic works, including sculpture, are naturally protected by a copyright that arises immediately in any expression of creativity. Software and data can also be protected by copyright. A copyright will often last for a period commencing from the date of creation or from the death of the creator. Development in recent years has seen other IP rights utilized to protect intangible assets such as topography rights for microchip designs and design rights for industrially applicable designs.

## Intangible assets risk and reward

There can be no doubt that the output of industry is increasingly conceptual rather than physical and that this is at last being reflected in the value of innovative companies. It has been estimated that around $1 trillion per year is invested by US private firms alone into intangible assets,[6] so this is not simply an area of concern for FTSE 100 firms alone, or for just the computing powerhouses. The wealth of assets at an organization's disposal are now likely to include more intangible assets than tangible ones, which should also shift the emphasis of risk and reward in the board-room and throughout the company. The considerable upside of these assets is clear as the innovator can muster a new line of financial reward from generating and maintaining assets other than its tangible property. A company's image and success can often drive a 'brand premium' above and beyond the usual market price for a product. A newly developed vacuum cleaner can reap significant revenues without the fear of generic copying because of the protection given to intangibles. Intangible assets can, in theory, be easily and cleanly transferred from one party to another and at greater speed, so allowing the creation of wealth at a fraction of the cost in a fraction of the time.

Intangible assets are unusual in that they can be a scarce commodity because ownership is granted as a monopoly, but once 'owned' they are easily deployed and can be used as many times as other resources will allow, and deployed at the same time elsewhere. The value of intangible assets increases exponentially depending upon use, because of the low transaction costs but also the initial ownership costs. Possibly the greatest upside risk, and one that is least understood in an organization, is that an intangible asset creates value for the future. This future value is probably ignored from a downside risk perspective because both brands and IP are considered a business risk closely tied to a product or service brought to market rather than broken down into identifiable elements that could be managed and controlled. This position isn't helped by the fact that being of future value, and by no means guaranteed, the accounting community has historically struggled to delineate them. Just as there are risks to tangible assets that are generally managed and controlled, there are significant legal, financial and indeed economic risks linked to intangible assets that must be considered in order to protect these main fiscal drivers.

Before looking at specific risk considerations for goodwill, brands and IP it is worth considering the characteristics of intangibles and how they may differ from a risk perspective from tangible assets. R&D expenditure is considered a measurable intangible asset by many and is also indicative of the attitude towards innovation. This process of innovation itself carries considerable risk of success or failure; the search for a superior innovation is by no means guaranteed. Any success may also be time-bound or commercially unexploitable; as the saying goes, 'Benjamin Franklin may have discovered electricity, but it was the man who invented the meter who made any money'.[7] This demand on innovation for profit growth's sake, and often unusually high profit, is an element of capitalism that carries a macro-economic risk according to some economists. In his seminal tome *Capitalism, Socialism and Democracy*,[8] Joseph Schumpeter describes his 'creative destruction' theory whereby an incessantly internally evolving economic structure incessantly destroys the old structure in creating another one. This theory can be readily applied to intangible assets, particularly IP innovations, where it must be questionable policy to innovate unless the savings made by the new innovation are substantial enough in quantum (presumably from a financial perspective in a profit-seeking organization) to cover not only the R&D investment poured into that innovation, but the investment not yet realized from pre-existing methods or technology. At any rate, from a global perspective it is doubtful that such innovation can lead to a Pareto improvement.

Reference to 'creative destruction' was common during the dotcom boom of the 1990s, in the corporate boardroom, as a supposedly old economy was replaced with the new. However, the internal and external risks inherent in such a new economy were infrequently examined during this period and often carelessly abandoned. Ironically the boom was soon followed by a crash as the entire structure was, indeed, found wanting.

The concept of risk from a macro-economic perspective has also been muddied by the incentive system for such IP rights as patents and copyrights, whereby an innovator can benefit from a non-strict period of monopoly for his or her creative developments. Though there are differences of opinion as to the economic effects of a solid IP rights system on innovation and economic welfare,[9] the prevalent attitudes are typically for a controlled monopoly right of such duration whereby losses at the margin due to a lack of competition are carefully balanced by the equivalent marginal gains from innovation.[10] This may be more easily said than done. Incrementally stronger IP protection may not necessarily lead to more R&D investment either, and could arguably prove detrimental to innovation if patent rights become limited. Interestingly some evidence shows that the strengthened IP protection provided under the 1988 Japanese Patent Law Reforms has done little for the development of innovation in the country but has increased the amount of patent litigation.[11]

The downside characteristics of intangibles lie in the difficulty to identify, manage and control them; this may be the very essence of risk management but intangibles are as yet still easy to shirk responsibility for mainly due to the need for support from other functions within an organization that lend expertise to this area. In some areas there is a limited ability to protect intangibles with ownership in the form of property rights. Similarly, if undefined, measurement and valuation can be difficult, or impossible, without expert advice and even then are often not considered appropriate to be notified in the organization's financial statements. As mentioned earlier, we are looking at future value where potential variables such as weak legal protection, replacement innovation, counterfeiting and many further specific risks can hinder exploitation. Additionally, despite potentially low transaction costs these may not always be realized because of misunderstandings about a non-physical entity. Such transactions can often be complicated, and complications increase costs. Where an asset has been derived from considerable sunk costs such as R&D and time, it is imperative to gain more than marginal benefit for an asset that might have an undeterminable life-span.

### External pressure to compose intangible asset risk management standards

'Everything that can be counted does not necessarily count; everything that counts cannot necessarily be counted', so said Albert Einstein. Taking the quote completely out of its context it presents a relevant comment on the relationship existing between accounting principles and intangible assets, if not now, certainly until a couple of years ago. Accounting value has always arisen from historical costs such as created stock, equipment or debt rather than future expectations. Intangible assets have therefore not been particularly well reflected through traditional reporting methods leaving something of an asymmetry of information for stakeholders in any corporate organization.

Internally created intangible assets have not been expressed at all, and the outlay to create such assets, such as the R&D or marketing costs, has simply been expensed rather than seen as going towards the creation of a valuable asset. Acquired intangibles have only recently been represented in financial reporting documents in any shape or form, ignored completely in merger accounting or lumped together and classified as goodwill during acquisitions. This has started to change with the implementation of new accounting standards for both the International Financial Reporting Standards (IFRS) and US GAAP.[12] Not only are acquired intangibles often now subject to identification, valuation and reporting, but there is also a breakdown of intangibles to separate assets with determinable lives, such as patents or copyrights, and those with undeterminable lives, such as brand, goodwill in its true sense or customer lists. The Accounting Standards Board of Japan (ASBJ) has historically lagged behind the West in its treatment of intangibles, but many have realized the disadvantage at which Japanese business will find itself if it does not account for all its assets. In January 2005 the International Accounting Standards Board (IASB) and the ASBJ agreed to launch a joint project to reduce differences between all of their accounting methods.

Presenting intangibles on financial statements, and effectively creating newly recorded assets, will have a considerable effect upon their treatment in the boardroom and beyond. Immediately there is something else to lose, and its risks must be considered. The increased transparency to stakeholders of such assets will ensure that they expect sound treatment of them. We are not yet in a position where all internally retained intangibles are listed on the balance sheet, but the impetus tends in this direction. Regardless of whether this actually occurs or not, the 'respect' for IP and brands can only improve. The penalty for not taking intangible asset values seriously on acquisition can have serious effects. In 2002, AOL Time Warner had to make a combined goodwill write-down of $98.7 billion because goodwill had been overpaid when AOL merged with Time Warner. Almost immediately the share price of the company had collapsed by 75 per cent.

International developments on the expectancy of corporate governance have also shifted intangible assets up the corporate agenda. As discussed by Ruth Taplin and Nick Schymyck in Chapter 1, the Turnbull Report proved a pinnacle to the UK's corporate governance movement of the 1990s and introduced the concept of a risk-based system of internal control. This encouraged the consideration of all material risks to an entity, and can be interpreted as including intangible asset value. Although traditionally laissez-faire in terms of interfering with big business, the US government implemented Sarbanes-Oxley in 2002. Developing three elements: governance, risk and compliance, Sarbanes has ensured their integration and placed a responsibility upon the board to implement to the benefit of stakeholders and sustain value in a company. Although intangibles are not specifically mentioned in Sarbanes, its combination with accounting requirements and the spirit of presenting an accurate financial picture of a company lends itself to this area. In light of the backdrop to the Act being the Enron/Andersen situation, this clearly also means securing reputation and brand image. Sarbanes also has a knock-on effect in Europe and Japan. Despite European businesses threatening to withdraw from US stock exchanges, due to the resentment of increasing compliance measures, any company with 300 or more US shareholders will be bound by Sarbanes-Oxley requirements.

The demise of Andersen demonstrated the disease-like qualities of reputation damage. Not only was Andersen, and indeed Enron, affected by the tarnished image, but it spread outwards leading business partners and customers to distance themselves from the problem. This illustrated that the risks to reputation and brands are not limited to problems from within.

There has definitely been, at least, a corporate governance trend of embracing risk from a 360 degree perspective, requiring strategies and plans for risk mitigation and control. Insurance, although not directly examined, also falls within the field of control. A growing appreciation of intangible assets, their values and their commensurate risks should not sit in isolation with the corporate board, but must be embedded at the grass roots and sustained throughout the organization. There is a long way to go before this point is reached. If risk managers are truly the stewards of risk, their impartial identification, analysis and control of all material threats to assets must be imposed upon all staff within their remit.

## Intellectual property risk

Risks to be considered by the risk management and insurance community, be they related to IP or brand and reputation, can be either of an internal or external nature. Insurance typically, although by no means always, lends itself to the concept of external risk because of the element of fortuity to an assured that is inherent in an insurable event. The risk manager whose duties go beyond that of insurance manager will have to be aware of all material risks to the business beyond those that have a recognized third-party transfer market. This section on IP risk and the following covering reputation risk will address both internal and external risks, but will naturally lean to the external and those capable of insurance for the interest of all readers.

From a general standpoint, IP risks can often be categorized into five areas as defined by KnowledgeLeader.[13] The first risk is that of 'availability'. Information that is considered IP must be made available to those who require it for the achievement of the organization's goals, be they staff or customers, but it must be protected from potential infringement. The second category is 'compliance' risk. There are many legal issues to be considered when dealing with IP, and this situation is not helped by the concept of IP meaning national rights only and so the law of many countries is often a factor. It is important to be aware of the legal implications surrounding IP. A third risk is that of 'brand', which falls within the IP context because of the relevance of trademarks and, to a certain degree, other IP rights. The protection of images and designs is associated with brand. 'Access' risk encapsulates the protection of trade secrets or know-how, and requires diligence in granting appropriate access to data or confidential information. The final category in KnowledgeLeader's quintuplet focuses on 'business value' risk. It is of paramount importance that IP assets are identified, valued and monitored to some degree so as to understand the value they bring to the company and to enable determination of changes in that value.

IP perils can cause a wide array of damage. Possibly the least damage comes from legal costs, penalties and damage awards following violations of laws and regulations, particularly with regard to civil actions of infringement or invalidity. There may be a significant loss of consumer confidence or respect impacting the bottom-line with regard to lower sales or profit margins. At worst, IP risks could realize a release of confidential information, loss of product line, brand or some other loss of competitive advantage. Such losses will certainly cause business interruption, and could effectively destroy the business. Samples of specific IP risks are now considered.

### *Patent risks*

One of the biggest risks to patents, and one that is best suited to risk management procedures and insurance, can be categorized as legal risk. As a category it certainly encompasses numerous perils to patent value. At its most base level, legal risk can incorporate product liability risk. Today's technology flirts on the borders of science and can consequently produce extreme results and side-effects. Although understandably considered from a product perspective, the relevance and separation of IP as a corporate asset will possibly require an understanding of loss of patent value from environmental, personal injury, property damage or other negligence claims. There is proximity between patented technologies and product liability in many areas, particularly the pharmaceutical and motor industries. Areas to be watched in the future might include nanotechnology, biotechnology and energy sectors.

The prospect of a patent invalidity or an infringement action is certainly a risk that must be considered. As with any legal action the risk can be considered from the standpoint of legal (and internal) expenses, liability damages and future business interruption. A patent action could possibly arise for a company either as a defendant or a plaintiff and costs are certainly going to accumulate when acting as either party. Jurisdictions vary dramatically in their legal costs, but one thing remains constant: IP litigation is invariably expensive requiring niche lawyers, experts, complex evidence and considerable time. The average cost of patent litigation in the US is around $2 million, with the UK and Japan around the $0.5–1 million mark. This is the cost of legal and court fees only. In comparison, employer's liability litigation costs typically only fall into the tens or possibly hundreds of thousands of dollars.

An invalidity action essentially involves one party attempting to remove the monopoly power of another by accusing the patent owner of holding a patent that is invalid for reasons such as existing prior art (an indication that the patent 'innovation' was already anticipated in the public domain), failure to disclose the best mode of practising the invention, or some sort of fraud during the patent application. It is important for those not familiar with the role of IP rights to consider that this action can bring to an abrupt end a product or service within the company. The loss of a core molecule patent, for example, can effectively mean open-season for all generic drug manufacturers to produce and sell the drug affected. R&D expenses may have been wasted and future revenues possibly limited to such a degree that the product line brings little or no profit. With patent offices around the globe persistently scrutinized for the quality of their patent grants, the practice of bringing invalidity actions has become an increasingly popular tactic.

An infringement action arises where one party alleges that the other is making, selling or using a product or service that is covered under the claims of a patent. This process can be particularly lengthy as there will be much legal discussion as to what contributes infringement of a given patent, and consideration as to whether there has been a literal infringement of the claims, or an infringement of the spirit of the patent. In the case of infringement actions there is also the cost of damages to consider as well as the legal costs and business interruption. Damage awards for infringing patents in the US have run into the hundreds of millions of dollars, with a doubling of awards or settlements exceeding $15 million in the last five years and 50 per cent of damage awards exceeding $5 million.[14] Europe and Japan have historically not handed down quite such large awards because of their different approaches to punitive damages, but awards are increasing and damages in the tens of millions of dollars are now not unusual.

Again, what must be considered is the effect a loss of rights perceived as owned by a company has upon its future performance. If a company is found to have infringed another's patent there are likely to have been large direct financial implications regarding legal costs and damages. In addition, to perform a service or produce a product in a way that was once done may now have been determined illegal. This ensures that the infringer must now find another way to provide such a product or service or indeed pull out of that market sector. To maintain a presence in that area may require huge expense on 'design-around' strategies, or the need to try to negotiate a licensing contract with the owner of the patent. There will be costs associated with obtaining the licence and also royalty and licensing costs incurred which can vary depending upon the industry sector and the importance of the patent to the product or service. Profit margins will certainly be damaged in one way or another. Small businesses will have their very livelihood threatened and often will not emerge from an IP dispute. Larger firms are certainly likely to see an effect upon their share price and a drop in margins.

Of topical interest is the position of employee inventors who generate IP that is assigned to their employer. Developments in Japan are proving to dispel traditional notions of selfless service to employers with large awards being granted to employees and ex-employees who have brought legal action over 'insufficient' awards for their inventions. In 2002, Masayoshi Naruse sued food and spice maker Ajinomoto for ¥2 billion ($18.7 million) for recognition of his contribution to developing a mass-production process for the artificial sweetener Aspartame. He was eventually awarded $1.75 million in 2004 by the Tokyo District Court. In 2003 Canon was in trouble with a $10 million suit brought against it by an ex-employee with regard to computer technology. In 2004 *Yonezawa* v. *Hitachi* resulted in an award of ¥160 million ($1.5 million) to an employee electrical engineer with regards to DVD technology. The Hitachi case was also particularly significant because the court ruled that compensation must be extended for patent applications in foreign territories as well as Japan.[15]

The years 2004 and 2005 have forced Japan to breaking point on this issue. Article 35 of The Japanese Patent Law takes on a new aspect with regard to 'reasonable remuneration'. In January 2005, former employee Shuji Nakamura agreed to settle with Nichia Corporation, over his role in the creation of the blue light-emitting diode, for ¥844 million ($8 million). The award is considerable and would have been some ¥20 billion ($187 million), had the Tokyo High Court upheld the findings of the District Court. Despite some relief that the District Court award was not upheld and the Japanese Business Federation even commenting that the settlement was 'sensible',[16] having only been originally awarded ¥20,000 ($190) by Nichia for his work Nakamura has sent shockwaves through Japanese industry. An estimated 43 per cent[17] of Japanese companies have now removed their upper limits on employee compensation, including Canon, Honda, Takeda Pharmaceutical, Sumitomo Pharmaceutical and Mazda. Western-based companies with R&D activities in Japan will of course also be subject to this issue, and in July 2004 an ex-employee of Pfizer Japan filed a ¥1 billion suit for hypertension drug technology.

The situation in Japan may be in dire stress at present, but the risk of employee or ex-employee actions are possible in other countries, if not in connection with compensation, then with ownership of patents where there is some element of doubt regarding the terms of employment or work done by employees 'away from the office'. Typically the situation is more certain than it is in Japan at present, with Western countries also historically better known for a strong 'incentivization' and compensation culture, offering stock options, pay reviews and promotions as well as financial bonuses. Concerned that Japanese R&D might diminish with recent decisions, an amendment to Article 35 has been rushed through to be implemented in April 2005.

Somewhat ironically the UK Patents Act 2004 has implemented a change in patent law concerning employee inventions in early 2005, which will effectively ease the ability of employees to bring an action. The UK has a provision under the Patents Act 1977 to allow employees to claim compensation where a patent has been of outstanding benefit to the employer. Since the Act was implemented, however, there has been no recorded suit on the matter. Many think this is due to the difficulty in showing that the outstanding invention was brought by the patent itself, rather than simply the invention in its broader sense. The new law will change this so that the employee need only show that the patent or the invention or both have been of outstanding benefit. Companies have every reason to dwell on recent employee legal actions and consider the material effects upon the business. A ruling of a 5 per cent 'royalty', or lump sum, to employees involved in the R&D of a product could have significant bottom line effects and slash profit margins.

Other patent risks that certainly fall outside current insurance schemes are the issues of expiration of patents, opening the market up to other players, or credit risk following a bankruptcy of a licensor or the poor record of a licensee. The expiration of patents can have devastating effects, demonstrated by pharmaceutical giant Merck's warning of a fall in 2002 profits due to the expiry of its blockbuster Prilosec. The anticipated loss of the best-selling prescription drug at that time led to a 9 per cent fall on the New York stock exchange in one day.

All the risks considered in this section may also have considerable third-party consequences, for example where patents are licensed to others and contractual terms provide for liquidated damages, or where such patents are used as collateral against loans or investments.

### Trademark risks

Due to the proximity of trademarks with the concept of brand, the issue of trademark risk will be explored in some greater detail later on in this chapter. However it is important to recognize some of the more basic risks at this point. Trademarks generally fall into two categories in most juris-dictions: registered and unregistered. Once granted, by a governmental body, a registered trademark is considered a valid entitlement to use and display the mark while also providing a property right that must be respected and so not infringed by a third party. The trademark owner will often seek to register a mark in order that it has a prima facie position of validity and to notify others of this. As a premium for this level of 'certainty' of validity, the trademark owner pays a fee to the relevant office and will retain the mark for as long as those fees are paid. It is vital to ensure that all trademarks are therefore managed and fees paid promptly to avoid loss of the right.

An unregistered system of trademarks may exist, such as in the UK where common law rather than statute law provides legal protection to those who have marks that are not registered but that have been imitated by a third party so as to cause confusion and pretend that the goods or services originate from the other. In such circumstances there is a burden of proof to establish that there is a recognizable mark being 'passed off', which is not necessary if a trademark is registered. Under both categories there is generally a burden upon the plaintiff to show that the defendant has directly infringed the plaintiff's use of the mark, or has caused a likelihood of confusion about the origins of the defendant's goods or services. The laws for trademarks are national, despite guidance under international treaties, but infringement typically revolves around the overall impression of an 'infringing' mark such as its look, phonetic character and meaning, the similarity of goods and services the 'mark' is applied to and any likelihood of actual confusion. Trademarks can also be lost through loss of enforcement or use, as discussed later in the chapter.

Trademark cases do not attract the same legal costs as patent cases because they usually require less evidence and expert testimony, or require less technical comprehension. This does not mean that they are cheap. There is also quite little likelihood of damage awards being made against infringers, somewhat noticeable in the usually litigious US, although the repercussions of a confused customer base or loss of a presumed owned mark can be significant.

### Copyright risks

For many businesses the risks associated with copyright are small. Due care might be required in promotional material to ensure that information is not plagiarized from others, and that fair credit and reference is presented. Additionally there is the risk that such material might be copied and infringed by a third party. Issues such as these are often considered mundane or trivial, with little risk attached or financial consequences. For those companies involved in the collation of data or in the creative arts industries, however, the copyright issues are considerably pertinent.

In this information age, the predominant IP right behind software is that of copyright. The US is considerably more liberal in awarding patents for computer software but Europe has been hesitant, awarding in some cases and not in others, with the law still not entirely certain and the future of software patenting still in open discussion by member states. Other predominant infringement risks associated with copyright have been demonstrated in the area of music downloads and the illegal copying of music and films. Databases held on modern computers can hold huge amounts of information and can be potential goldmines for an infringer. Due care must be given to protect such information and to retain authorship of it. There has been substantial counterfeiting activity in the software and film arena with China and Russia among those responsible for a considerable amount of copyright infringement. Copyright actions are, again, not as expensive to litigate as patent actions, but the impact of the associated risks or damage awards can be considerable. As with employee inventors, much of the litigation surrounding copyright arises from contractual disputes over ownership or use of copyrights, or the terms of royalty payments.

### *Trade secrets and know-how*

IP such as this is defined as all that intellectual capital which is retained within an organization and yet is unregistrable or has been deliberately retained as exclusive knowledge for the benefit of the owner. Such IP might be retained within any given functional division of an organiza-tion, be it marketing, production or otherwise. The information should be treated as an asset just like any other intangible and the issues of 'avail-ability' risk and 'accessibility' risk are of particular relevance. Choice and monitoring of staff, levels of access to information and carefully drafted contracts with internal and external parties should be top of the agenda. Often the biggest risk with trade secrets and know-how is failure to exploit it to its full potential to give the organization a positional advantage among its competitors. This can be addressed with carefully defined internal struc-tures and information exchange between divisions where possible. One company that has built an empire around its trade secret (and brand) is Coca-Cola. The recipe for Coke is known by only a handful of employees in the entire company, and is retained as a trade secret rather than any other IP right so that production details do not have to be presented publicly in a register.

## Reputation risk

Much has been written on the subject of reputation risk, although it is possibly one of the least comprehended within industry, or at least has had a minimum amount of resources and man-hours poured into it at a risk management level. Likewise, the insurance industry (and other potential third-party transfer players such as the capital markets) has not driven the initiative in this regard but has waited for a consumer demand that has not been forthcoming. With both sides often failing fully to quan-tify and understand this most intangible of concepts, little has been done. The risk management community has expressed its fear of reputation risk without taking the plunge to purchase any form of insurance cover.

Of all the intangible assets, reputation and brand are probably the hardest of all to grasp in terms of insurability. This does not mean that it is impossible, but that time and effort needs to be taken to develop cover. The future of reputation insurance is not helped by the fact that insurance capacity will be hard stretched to cater for larger organizations. The effects of a reputation fall-out can be disastrous, with often huge drops in share capital and/or threats to the ongoing viability of the entity. For the more sizeable companies, i.e. those with market capitalization much bigger than the largest insurance and re-insurance companies, the potential for providing meaningful insurance is debatable.

Managing reputation risk goes way beyond the risk manager's role, but the risk manager will be integral to the overall strategy and system generated by the company. As the guardian of risk, the position requires an understanding of the issues and the potential role for insurance. The more pro-active risk manager will seek to implement and control the activities of others in the organization, reporting through the chain of command any areas of weakness, and suggesting procedure and practice to help minimize and mitigate risk. The following section comments on some of the issues at hand.

### The intellectual property crossover

As mentioned earlier there is an overlap between IP and goodwill, particularly with regard to the intellectual capital of firms held as trade secrets and know-how and the branding element, which is generated partially through the rights of trademarks. The risks associated with these areas must be considered due to their potentially exponential financial effect upon the reputation and brand image of the organization. The loss of customer lists, or advantageous marketing data, can enable competitors to take market share. The loss of key individuals with production knowledge or creative input could also damage the future earnings of the company and demonstrates to stakeholders that future performance may not be as good as expected.

Trademarks are particularly closely associated with brand. Although brand image amounts to many more things than a sign, it is through trademarks that origin and association to products and services are derived. The sign or symbol will come to reflect all that is good or bad about the owner in the mind of the customer, so much so that the right mark on a given product can generate a brand 'premium' at or above the usual rate for a similar product or encourage further sales than would have occurred without the 'label'. The quality of goods may in many cases become a secondary requirement to the buyer, as the perceived qualities of the brand are instead instilled.

Trademarks require care and attention to protect the brand of a company or a product line. Even if a trademark is gained for a product or company name, it must have its integrity preserved as a source identifier rather than a generic reference to a product of that sort. In other words the value of a trademark, in the sense of its exclusivity and as a symbol of origin, may be lost if a word incorporated within a trademark is taken on as common usage to identify a category of product and therefore used as a verb or noun. The trademark itself is not necessarily lost as a registrable right but it can become difficult to prevent other suppliers of a similar product using the term. Not only would the gained intangible assets be lost, but further expense might be required to develop another. This is probably best explained by way of examples. In 2002 Sony rather publicly lost the right to its registered 'Walkman' trademark used on personal music cassette-players in Austria. Arguably the value of the trademark had been lost even some time before as for a number of years the term could be found in dictionaries. The Supreme Court ruled in this case that the term 'Walkman' had become generic for such technology.

The threat of diminished value of trademarks, popularly termed 'genericide' is not a new threat, but one that has created a number of verbs and nouns in the common parlance, where origin is rarely, if ever, considered. The words aspirin, escalator, thermos, trampoline, gramophone, cellophane and linoleum were all once protected trademarks that have since lost their origin identifier and are now known to identify a whole class of similar products. Ironically it is often the more unusual and distinctive names and technologies that lose out because of their creation of a product line that had no former description and were particularly novel. The brand name became synonymous with the product, which was later copied or designed around by others. In recent years this can be seen with 'Xerox', who spend millions of dollars in marketing to notify the public, and the legal system, that 'Xerox' is a product and company brand rather than a generic term for photocopying. 'Hoover' has struggled for many years to retain the name as a brand rather than an alternative term to vacuum-cleaner. Going forward, new technologies will generate such problems, and there must be awareness of the problem. 'Google' in its IPO filing states, 'there is a risk that the word Google could become so synonymous with the word search. If this happens we could lose protection for this trademark'.[18]

Another risk to trademark value, and indeed many other forms of IP, is dilution as a result of a poor licensing strategy. This area will be touched upon later in the chapter, but is certainly relevant here where there is an IP and brand crossover. The problem with regard to trademarks exists particularly in the domain of luxury goods, where there is not significant control of the 'franchise' leading to a potential decrease in exclusivity of a brand and/or a decrease in the qualitative standard of the goods or services. The brand 'premium' can be lost if, in the eyes of the core consumer base, the distinguishing features of exclusivity and fine quality are lost due to the attempts to gain additional market share or profit. This is an area that should be closely controlled by the owner of the rights, and does not fit comfortably within the parameters of insurable risk.

### The concepts of pure brand and reputation risk

There have been limited attempts to define brand risk in risk management circles, and generally the concept is treated too narrowly to catch only those particular perils that the risk community, rather than compliance, legal or marketing division, can feel reasonably comfortable with. In fact brand

and reputation risk should broadly consider all stakeholder perceptions of an organization that might threaten the sustainability of that entity in the short, medium and long terms. Too often the concept falls short of this and incorporates at best the notion of business interruption following other more familiar and insurable perils such as product liability lawsuits, product recalls or other such risks. Extending the theory of the brand 'premium' into the brand risk concept, of course, requires that all threats to consumable differentiators are included, but the most elementary risks to the operation of all similar organizations, such as ethics or health and safety, must not be forgotten in this process.

With regard to brand and reputation risk, prevention is certainly better than the cure because of the cataclysmic and non-linear effects of a loss. A brand or reputation can provide 60 to 80 per cent of goodwill assets to a company because that wealth is supported by every transaction and interaction the company has with its stakeholders. If reputation serves as currency for each participation with third parties then one threat to that repute can have dramatic domino effects upon all stakeholders. Losing a good name in the short term will usually cause considerable financial damage (depending upon the competitive nature of the market and the availability of choice for a consumer) in the form of lost profits. If substantial harm occurs, the damage may be irreparable, causing diminished market capitalization and shareholder value or loss of market share and consumer base. It has become a public relations cliché that building a reputation from scratch is far easier than restoring a damaged one.

Of course a reputation may be damaged from within and without an organization. Gerald Ratner describing his high street jewellery products as 'crap', leading to the downfall of Ratner's, is clearly self-inflicted damage in the same way as a product liability issue. When a business partner suffers a loss of some kind, whether reputation or not, a reputation loss can still fall upon the company that had little or no control of the other. Today's economy depends upon millions of interwoven relationships, whereby a product or service branded from one party is often formed using the skill-sets of others. Association with a party subject to a damning ethical, corporate governance or liability reputation loss will spread like a disease to all who come into contact with it. Never has this been seen clearer than in the Andersen situation, where not only was the accounting firm to suffer from its tarnished image, but aspersions were cast upon other companies that hired them. Reputation threats move in a way that is quick and deadly, ensuring stakeholders cannot afford to be associated with those companies.

Allegations can be as harmful, especially in the short term, as any genuine reported failing within an organization. The media and public are unforgiving and risk must be considered from the perspective of the effect rather than simply focusing on a cause. Despite the analogy of there being 'no smoke without fire', it is possible to suffer severe disruption without an identifiable 'peril'. In 1999 Coca-Cola was put to the sword by the Belgian government, which demanded that 30 million cans and bottles of soft drink be recalled following news that up to 100 people had suffered nausea and cramps after drinking the products. The Coke management withdrew the products and even found an explanation for the possible sickness at the time. However, what was barely published in the media was the fact that, after the scientific study of the product, there was nothing in the product capable of producing such poisonous reactions. Coke's reputation was damaged by what many experts now believe was a form of contagious anxiety, heightened by the recent local dioxin scares in animal feed and chickens. The costs to be borne extended well beyond the recall expense in the manner of business interruption costs and impaired future earnings.

The treatment of reputation risk from a broad perspective will lead the risk management profession to incorporate all manner of threats to stake-holder perceptions of the company and, as a result, the future sustainability of the organization. Identification under such a premise can generate an array of perils such as catastrophe, recall, product liability, treatment of employees, corporate governance, fraud, business practice, ethical choice, customer service and health and safety among others. Companies that are particularly vulnerable to loss are service industries, which by their nature hold most of their assets as intangible goodwill and intellectual capital. Consultancies, charities, financial services and purveyors of high street brands should certainly consider their dependent relationship with consumers.

## Intellectual property transaction risk

With the increasing appreciation of IP rights, particularly in high R&D spending or high-tech industry areas, the intangible assets of a business now find themselves subject to treatment in the market as a valuable, tradable commodity. We have already touched upon the role of licensing agreements to generate revenue streams in areas where a business may not have knowledge or access to a certain market, or where IP would otherwise remain dormant if it is not a core technology to the business. The licences may be non-exclusive (in Japan, *tsujou-jisshiken*) or exclu-sive (*dokusenteki-tsujou-jisshiken* or *senyou-jisshiken*) depending on the needs of the respective parties. These arrangements can reap considerable rewards. IBM is a case in point, having led the initiative in the 1990s and currently generating licensing income of over $1.5 billion, an estimated 80 per cent of which goes straight to the bottom line.

In the future we can expect that the wealth associated with intangibles will be unlocked to a greater degree through mergers and acquisitions (M&A), IP sales and a more forthcoming approach to using intangibles as collateral from the financial markets. The accounting developments mentioned earlier will help and not hinder this progress, but will also highlight the values at stake and motivate further risk analysis. Although intangible assets are of tremendous value to the company today, their increasing appreciation by other competitive and financial markets will drive the relevance of due diligence by both transacting parties.

Traditionally the preserve of 'taking out' competition and increasing market share, M&A activity is increasingly motivated by objectives of incorporating intangible assets. IP may be a means to access a brand new market, save design-around costs or to ensure freedom to operate in an existing market place by obtaining 'blocking' IP. Although the goodwill or reputation of the entities combining may be critical in the success of a deal, it is the knowledge-based assets such as IP that are the new economic drivers. Risk managers may not have historically been much involved in looking at IP acquisitions, but their risk identification skills and knowledge of the insurance market should mean that they are a critical component in any due diligence team in addition to the lawyers.

In its purest form, IP due diligence is a systematic attempt to expose, document, evaluate and perhaps value the relevant portfolio exchange in the deal process. With this information the acquirer should have an improved understanding of the target company, any untapped potential revenue exploitation and, of course, those significant downside risks such as potential liabilities or weaknesses in the IP portfolio. With this information the deal may face considerable re-negotiation, depending upon the motives for the exchange, as certain risks may represent additional costs to the deal. 'Reps and Warranty' insurance protections have been common for a number of years and there has been an increasing development of both legal expense and liability policies for IP risks, followed by genuine first-party IP value insurance should depreciation occur after the deal. The same issues are also directly relevant to companies going into a public offering on the stock exchange where perceived value is retained in its proprietary technology or brand. In this instance, IP value and risk will need to be understood by analysts and prospective buyers.

Businesses may increasingly find that IP perils have more financial impact than simply a loss of revenue stream, or enforced licensing arrangements. There is increasingly a tendency to utilize IP as collateral in loan arrangements as financial institutions look for other lines of income, and technology-rich companies try to raise cash against their predominant assets. Such IP monetization is a great step forward and underlines the increased respect given to intangibles, but again requires great due diligence by the lender, and an appreciation by the borrower that their own IP risks could ultimately lead to default on loans. Default could mean exceptional penalty terms, a short-term refinancing arrangement or even failure of the business if cash support is withdrawn. The financial institution will, of course, be concerned about loss of any sunken cash and the asset value of the collateral should elements of it be lost or unexploitable.

In much the same way, the risks inherent in IP monetization can be found in the area of IP-backed securitization which continues to gain momentum in the field of asset-backed securities. Estimates on the amount of business in this area show that in 1997 there was a meagre $380 million raised in IP-backed securitization, yet this figure had risen to nearly $15 billion for 2002.[19] Following on from initial transactions, such as the 'Bowie Bond', which raised cash for musician David Bowie, deals which pool future potential revenue streams in an instrument that can enable a rating agency to assign a security rating that will attract a different class of financial markets to IP have proved a popular alternative to standard debt-like securities. However, as the Bowie Bond has shown in recent years, the risks associated with IP can seriously damage the rating where IP revenues are hindered. In this case the popularity of the recording artist at certain times and the damage caused by illegal downloading and pirate copying activities can seriously affect the value of copyrights. Businesses must be aware of the impact of intangible asset risk upon their financial relationships and investments.

## Current intangible asset risk management practice

Despite 95 per cent[20] of investors and analysts believing that there must be more than just ownership of IP, meaning protection and exploitation of it as well, there is considerable evidence to suggest that current practice does not embrace such a wholehearted management of intangible assets. This approach is endemic to many companies from board level down through to risk managers and legal teams and must be improved in the future. A survey commissioned in 2004 by law firm DLA[21] found that European businesses are failing in their management of IP with less than 40 per cent of IP-owning businesses knowing even the value of their intangible assets. Only 52 per cent of businesses had a documented IP strategy with a similarly small percentage of management boards retaining a representative for matters relating to brands, IP or R&D. Considering that innovation should bring considerable upside risk through the production of IP rights, it is sad to see that the same survey projected that about £8 billion of income is lost from failing to maximize IP value in Europe alone. Other survey evidence is supportive of this position with one survey of senior executives finding that only 5 per cent believe their company has a robust system that measures and tracks all aspects of intangible assets and intellectual capital.[22]

There is no doubt that sound risk management of intangible assets should be complementary to an overall IP-friendly business strategy. Intangible assets do not yet have a respected position as a strategic business risk and the considerable R&D investment is too often not matched against the asset value it returns. When the values are finally realized the concept of risk to those assets will surely improve. Regardless of the company's overall strategy towards intangible assets the individual disciplines within the organization, such as the legal, accounting and risk management divisions, should still be aware of their professional duties and provide feedback and industry standards to the board. There is some difficulty for these areas of expertise to enforce their findings through the organization though because of the usual 'cost centre' mentality that is forced upon them. This approach often leaves these divisions either unable to venture into exploitation of IP or with a budget parameter that prevents allocation of resources and the consideration of elements beyond the traditional job specification.

The risk management department must, however, try as best it can to collaborate with all divisions in addressing all risks to intangibles, and consider the best methods of identifying, managing and controlling that risk. Just as the risk manager retains duties over property and plant risks despite the presence of real-estate managers, there can be no excuse that intangible property rights fall under the jurisdiction of the legal department and not within the risk management remit. Risk is indiscriminate and must be dealt with regardless of where it might occur and whether it falls within the risk managers' comfort zone. Too much is at stake for the company. If certain risk management functions are best performed outside the conventional disciplines, then this must be monitored by the risk management function with expertise freely given where relevant and certain responsibilities, such as insurance knowledge, retained.

No doubt intangible assets have an abstract feel about them that deters many from embarking on a thorough investigation. Avoiding confrontation with the issues, although an easy way out where upper management is in a knowledge vacuum, cannot be the way forward for anyone in an organization, be they risk management, legal, accounting or compliance. Often the potential gravity of these abstract risks to an organization is anticipated, but there is an absence of a pro-active approach to identification and control, and a structured strategy is non-existent. Surveys by AIRMIC, the UK's Association of Insurance and Risk Managers, and other insurance bodies have highlighted the growing fears for business. Until recent years the risk of fire was still the main cause of concern, but the typical top three greatest risks as perceived by companies themselves are more likely to be viewed as:

1   business interruption;
2   loss of reputation;
3   products liability/tamper/brand protection.[23]

Aon, the global insurance broker, in 2003 produced further survey evidence that showed business interruption risk to be the greatest threat perceived by companies to their operation, yet the elements of this business interruption are often interwoven with intangible asset perils. Despite 'loss of reputation', noted as the fifth highest major risk (having been first in Aon's 2001 survey), it would appear that only 34 per cent have a risk control plan or strategy in place, and a mere 28 per cent have undertaken a risk audit to consider their individual brand and reputation risks.[24] While there are signs of improvement from previous surveys, there is still a long way to go until these risks are given full consideration.

Ironically, the achievements coming from corporate governance standards and stakeholder pressure over the last few years may have taken the responsibility of managing day-to-day risk away from the professionals who know most about it. It would appear that there is a growing trend to lay ownership of responsibility for major business risks at the door of the executive board rather than any risk management department. Survey evidence suggests that just under 90 per cent of companies retain this responsibility at a board level rather than purely at department level.[25] Mitigation strategies and the measurement of the effectiveness of controls are also generally retained at this higher level and there is a reduced role for risk management departments. The pressures on boards to assume more responsibility could possibly be detrimental to the company should the strategies emanating from it not have benefited from considerable thought and cross-company buy-in. The Aon survey also suggests that the number of companies with risk management or insurance departments has dropped radically from 86 per cent to 54 per cent, and, if this is true, reflects a quite shocking position.

## Intangible asset audit and risk management

As the old Chinese proverb states, insanity is doing the same thing in the same way and expecting a different outcome. Eventually both the sensationalized and not so high-profile business failures and financial losses attributable to intangible asset risks will force the hand of companies to pay more attention to such issues. It may take a direct loss to a company before it realizes what it truly holds at stake. This is often the case, particularly in areas that are tricky to comprehend or measure and that are thought of as 'new'. The matter is not helped by the fact that issues such as loss of IP rights are often not front page news and companies that have suffered loss quite often prefer to wash their dirty laundry indoors. This often leaves their peers unaware of the consequences, and unable to determine the best risk management approach. Of course, over 70 per cent of companies that have a major business interruption will not survive to tell the tale and so need not align their risk management strategy in the future.

The most obvious statement to be made about understanding risk in the organization is that risk managers or board executives must know their company. They must understand where the revenue drivers lie and what privileges those drivers enjoy to maintain their success. When analysing risk, attention must be paid to the entity and not to the textbook, giving an unbiased account of all risks to tangible and intangible assets regardless of whether the same risks are identified by peers in the market or represented in the insurance industry. Identification might need the cooperation of other personnel in the company. Managing and controlling any intangible asset risks certainly will. The types of potential loss, the scale of those losses and likelihood of those losses must all be considered and expert advice will often be required when dealing with intangible assets. As with other areas audit, planning, prevention, communication and response are all core principles with mitigation strategies simulated, rehearsed and benchmarked. Of course, it is also crucial to ensure with intangible assets that opportunity is maximized.

The Risk Management Standard, published by the Institute of Risk Management in conjunction with the Association of Insurance and Risk Managers and Alarm, sets out a list of duties that includes knowing how the organization will manage in a crisis and ensuring that there are appropriate levels of awareness throughout the entity. There is typically a considerable lack of this within most companies, particularly in small and medium-sized enterprises (SMEs). The effects of a legal claim against IP, loss of confidential information, or a stinging piece of journalism must be considered from a financial, strategic and operational standpoint, and each element might require its own contingency plan. At the basest level, what must be assumed throughout the company is an awareness of the existence and nature of intangible assets, their value to the organization and any current management of them.

The strength of the brand and a reliance on reputation must be considered, and again this should not be limited to a benchmarking against others who may well have a higher dependency yet not have implemented an audit or plan. The relationship with third parties will be integral to this analysis as well as the competitive market place. The interface with stakeholders must also be examined, such as prior relationships with the media, if any, or the ability to understand and deal with the media following a potential problem. The values at stake must be considered and this might require some valuation to – if not fix firm numbers against the assets – at least alert relevant divisions to the brand premium obtained because of a position that many take for granted.

The threats to an entity's reputation are numerous and increase over time as corporate governance and 'regulatory capture' highlight what is and is not acceptable. In addition to this, society is becoming all the more informed through the powers of the media. Society has over time been empowered in many areas to voice its concerns and vote with its feet, be that at national elections to demand regulation or on moral and ethical matters. More choice has been given to the consumer and this heightens the scale of potential risk. Organizations must take stock of their individual situations and appreciate that a product recall could cause a loss of reputation with immediate and ongoing financial loss (and which is not covered under a standard recall insurance policy) as could a quite different media report on business efficacy or financial irregularities. It is important to work closely with initiatives within the company that highlight the elements of quality in the business or that provide social support so that in dire need these campaigns can top up the image of an entity and engender goodwill.

Incident response will need to be swift with decision-making and empowerment clarified as well as optimal allocation of resources to mitigate and control. There must be a business continuity plan that flows from any event and that can anticipate any required change of business direction be it in advertising, product launches or even any required acquisitions or immediate disposals. The framework must be regularly updated and tested.

There has been a movement in recent years of IP management that certainly comes from the perspective of maximizing gains from IP as much as analysing risk. IP management lends itself to those who have the greater understanding of IP legal rights, but also requires a commercial approach to exploit value from IP portfolios. There will be some need for communication between IP managers and risk managers, but risk must actually be considered by both parties. IP portfolio managers will look to focus resources on the best opportunities for the business in utilizing the IP and will identify patent clusters and prioritize non-core IP. There may then be further opportunities to generate revenue from non-core IP through licensing arrangements or potential tax breaks from IP donation.

IP risk management should utilize the tools and process described above to consider the organization's dependency upon IP. All of the company's IP rights should be identified and filed as inventory as should all IP that is not owned by the business but licensed from third parties. This stage might also consider whether the IP is being captured from R&D correctly and exploited to its fullest extent. The registration of trademarks and patents in areas where the business operates or where revenue could be drawn is also prudent, with considerable risk to the business faced should this not occur and freedom to operate or market share lost. The impact of loss of IP rights, potential damages to infringed third parties, design-around costs or licensing fees should all be considered.

Those taking responsibility for risk must instil awareness of IP into daily operations, from ensuring board-level involvement in the management of IP, attention to contractual provisions for indemnification of IP infringement or to regulate the handling of confidential information within the company. It is going to be important to value the IP retained by the company to comprehend fully the assets at risk, and a clear reporting procedure should be introduced for loss of IP and any IP infringements by or against the company. Records should be maintained of all employees involved in the creation of IP as well as all plans, drawings, lab notes and materials involved in the original creations.

After considering all of the options of avoidance and internal control, it should be considered whether the financial risk of intangible asset perils can be shared with a third party through risk transfer. Knowledge of the relevant insurance products or those insurance entities that will look to create new risk solutions is critical, particularly considering that the area is still relatively niche. A comment on the availability of insurance for intangibles is considered in the final section below.

## The role of intangible asset insurance

In the process of identifying, analysing and controlling risks to the business, the final stage will always be to consider whether the residual risk remaining, when all other means to minimize or eliminate have been exhausted, can be transferred to a third party. As mentioned in Chapter 1, the last five years of the twentieth century was a time when there was plentiful insurance capacity in the market which drove down premiums, improved policy terms for buyers and was characteristically a market conducive to purchasing a wide array of insurance covers. This period both helped and hindered the treatment of intangible assets within the insurance and risk management sector. The benefits, as will be discussed later, were that insurers were particularly interested in creating new products in search of other revenue streams in a 'soft', i.e. depressed, market. In addition the low premiums seen in traditional property and casualty markets allowed buyers' budgets to stretch to other niche areas of cover, such as intangible assets. This four- or five-year period therefore saw intangible asset cover grow into an insurance class of potential and was also in sync with the growing appreciation of the asset class by high-tech companies at least. From a risk management standpoint, however, the low cost of insurance often led to companies spending fewer resources on identifying and controlling insurable risks within the company because of the broad insurance cover that was available. The sudden hardening of the market following the events of 9/11 was to have a dramatic impact on all of these developments.

### Reputation insurance

The availability of reputation insurance has always lagged behind that of its cousin, IP insurance. The insurance industry has found it difficult to deal with conceptually because of the vague nature of the beast and insurance typically requires clear and defined insured values and perils. There has not, until recently, been much attempt by purchasers to define and quantify their own goodwill or to identify those risks that would seem most likely to prove harmful to the company's reputation. Without this initial practice by industry it will prove difficult for insurers to fill the void. Additionally, as mentioned earlier in the chapter, insurance capacity effectively to underwrite such potential losses may not exist for some of the larger multinationals. With market capitalization far exceeding some of the largest global re-insurers, the effect of a sizeable dip in share price or revenue may be too difficult for insurers to bear. The practical benefit of insurance must also be considered for companies should they effectively suffer a terminal reputation hit. Insurance proceeds might simply be sucked in by the assured organization as it spirals uncontrollably, unable to resurrect its business.

Although risk avoidance is clearly best practice in this area, there is still some potential for the insurance industry to help the struggling and not the terminal sufferers. The first efforts in this area were inclusions or extensions to standard product contamination or product recall policies that provided cover for fees and expenses to engage experts to deal with such issues. The intention was to ensure best practice and attempt to minimize business interruption and enable some crisis management. Eventually this concept was developed into a stand-alone insurance product by American Insurance Group (AIG) known as 'Crisis Containment' cover. The policy covered the fees and expenses of professional service experts required until a crisis was controlled for a maximum of 30 days or the insured limit exhausted. The crisis events were 16 named perils including, interestingly, loss of IP rights, restatement of financial statements or a negative sales or earnings statement, insolvency of a subsidiary, man-made disasters, fraud and, of course, recall. Though limited to the perils pre-determined by AIG, this cover was an interesting development in an area that had little prior attention. Unfortunately time has tested the cover and it is now not widely advertised as a stand-alone product.

There has been some further development emanating from Lloyd's of London with regard to reputation risk insurance. A policy providing first-party cover for loss following an adverse media report is still available in the Lloyd's market and is generally used to protect revenue streams and profit derived by an organization or one of its particular brands or product lines. Being first-party protection, it does not look to provide fee and expenses cover, but rather direct financial loss as a result of a report in the public domain that may have been based on truth or lies. An advantage over the pre-specified perils approach in products such as that of AIG's is that the product can be tailored to the particular concerns of the organization rather than buying a standardized product. Adverse media report perils will, however, be scheduled in the policy once the insurer and the assured have agreed with the scope of cover. By sitting the underwriter down with the risk manager of a company, the 'worst nightmares' of the specific organization can be offered some risk transfer. Perils can cover similar events as noted with other products and might range from the obvious recall (extending current standard Product Recall cover), to many varied allegations of unethical behaviour. The product has been of particular interest to service companies and charities whose reliance on a good reputation is high and potentially fragile.

Market capacity for such products has yet to reach the hundreds of millions of dollars level, but seems to have improved in recent years despite a difficult market for niche products. For SMEs with a heavy reliance on brands and reputation to the larger company with a significant strategic product release, the value of this cover is still clear and one could expect considerable growth here in the years to come. For products daring to tread new ground in the field of intangible assets, Lloyd's of London remains a significant and innovative market.

### Intellectual property insurance

The area of IP insurance is considerably more developed than reputation insurance but has suffered somewhat in recent years. Underwriting IP business had occurred on a stand-alone basis since the 1980s when insurers first realized that its lack of exclusion from general liability policies and incidental inclusion within 'advertising injury' claims was proving characteristically expensive. This type of cover related to legal expenses insurance, either as an enforcement or defence policy, with an element of liability for damages following an unsuccessfully defended infringement action.

The 1990s saw an increase in IP business written because of the softer market conditions and a growing awareness of IP rights and their value. The Lloyd's market participation also increased significantly due to its greater creativity in the area and some high-profile claims to US carriers. When US insurer loss ratios started running around 4,000 per cent or more, they retrenched a little and allowed US business to come to London. However, in the last quarter of 2001 everything seemed to change. The effect of 9/11 was to cast the global insurance market into a rate-hiking frenzy in all classes of business and the industry had the excuse it needed to return to a 'sensible' underwriting methodology. Although sensible underwriting was not to change the underwriting approach to IP directly, insurers typically focused on low-risk, high-earning business that has a short or limited period of liability and was not particularly specialized. However, capacity to write a wider range of liabilities is again increasing as capital is moved from less-profitable classes. As a result, and with the expected rate softening in property/casualty premiums in coming years, insurers are increasingly interested in more specialized lines of cover.

At present, however, there are a limited number of participants in the IP market. Although some reasonable profits have been made by underwriters from a defensive legal expense and liability standpoint many are reluctant to enter the market. IP rights are particularly volatile and not all companies rely on them to the same degree leading to some moral hazard and selection-against issues where only companies carrying higher risk seek IP insurance. Legal expenses and damages are also extremely difficult to quantify at the inception of a policy or even at the notification of a claim because of the complex legal and competitive environment that IP rights are in. This is not helped by the fact that national laws on IP vary considerably and there has not yet been significant empirical data to calculate insurance risk with regard to a range of rights such as patents, trademarks and copyrights. With such calculation proving difficult, some underwriters have looked to increase premiums above their true or competitive rate. This has further ensured that an adverse selection occurs whereby the policy is only worth buying by those with a considerably increased risk. It can become a vicious circle.

Much more analysis needs to be done to measure the magnitude of loss and the rate of infringement. This has proved difficult not only because of the comparatively low volume of policies sold from which to draw conclusions, but also because these elements cannot be assessed purely by the number of disputes that reach court. Many IP challenges are settled behind closed doors with little information in the public domain. Additionally many owners of IP rights have to give up a legal fight because the court costs are prohibitive. It is not clear what a predominance of IP legal expense insurance might do to the risk of infringement or to the number of challenges, although it should reduce the number of cynical infringements. In the last ten years a little more research has been done with some potentially valuable studies conducted in academia on litigation rates. Such studies have reduced the likelihood of litigation in the patent field by industry sector, age of patent, portfolio size and other characteristics.[26] Judgments have also been analysed from US trial data as have IP trial durations.

There is, however, the likelihood that the volatility and, indeed, size of risk could be lessened in the next few years due to a number of factors such as increasing corporate social responsibility and corporate governance as mentioned earlier in the chapter. These should have a positive effect on the rate of wilful infringement which, although underwriters try to exclude it, remains difficult. To a lesser degree national and European initiatives could also help. In the UK the new Streamlined Procedure will attempt to speed up smaller law suits and attempt to make the legal system a more cost-effective route for IP disagreements. Similarly, the UK Patent Office's new non-binding opinion on validity and infringement could also help settle claims earlier without undue legal cost or the roller-coaster of uncertainty that occurs during IP trials.

The availability of IP insurance must also be more widely advertised to the many small and medium-sized enterprises. At present there is not a large enough pool of IP assureds from which to calculate an accurate premium. A survey published in 2003 indicated that only 49 per cent of SMEs were aware of infringement insurance and 57 per cent were aware of enforcement insurance.[27] This has been a considerable problem for underwriters of IP business who are also typically dependent upon broker networks to spread the word on product availability. In recent years the larger insurance broking houses have shown little interest in IP and often do not even seem to give it a passing mention to clients who rely heavily on IP assets. The brokers have, arguably, in the hard market been able to place well-paying business in standard property and casualty business with less effort than is required for niche products. It is still an area that few brokers understand well and so is rarely promoted to clients.

The IP insurance needs of a company must be carefully considered on an annual basis. There must be consideration of whether certain IP protection is already afforded in broad professional indemnity or commercial legal expenses policies. Generally the IP element is excluded, particularly patent cover, but there may still be some reasonable if not extensive cover available. Of course, IP risks can be felt increasingly by directors and officers as well as the company itself and thought should be given to this factor. It is increasingly apparent that directors and officers in the US can be found liable for mismanagement of IP, examples being the extent of foreign patent filings, lack of valuation and disclosure of patent portfolio or an absence of disclosure of patent expiry dates to shareholders in high-earning products.

The stalwart IP cover, of course, relates to legal expenses and liability cover. The main markets are to be found in the US and the UK, although there are providers in other countries. An enforcement or pursuit policy will pay all the professional fees and expenses necessary in bringing an action against those who infringe the assured's IP rights. The cover can be used for many forms of IP right, and cover is often given on a worldwide basis, despite differences in IP laws, if required. This form of cover is typically considered as a cover for the David against Goliath situation where large companies cynically infringe IP rights of smaller companies because they believe they will not be challenged in court due to the excessive costs involved. The cover can sometimes be purchased on a 'known' basis – where an infringing third party is known at inception of the policy and a fighting fund is required – or on a 'no-known' basis.

The defence legal expenses cover will pay the professional fees and expenses incurred defending a claim from a third party that the assured's products or processes infringe IP rights of another. A liability element for damages or agreed settlements is often included and, with awards reaching astronomic levels, companies of all sizes may be interested in this protection. Additionally, defence cover can be formulated to pay all costs associated with defending a challenge to the ownership or validity of IP rights. Such cover can also provide protection in the cases of post grant opposition actions in the European Patent Office or Japanese Patent Office or interference and re-examination actions in the US. Variations on this form of cover can sometimes be found that act in much the same way that representations and warranties cover for indemnity clauses in contracts.

Although currently only offered by the private sector, there has been a great interest in legal expenses cover as a mechanism to assist in the enforcement of IP rights and to meet any failings of the current patent or trademark offices that initially grant the rights. This mechanism has been discussed at a national and European level, where there has been talk of some form of centralized scheme. A scheme was attempted in the late 1980s in France, but failed mainly because the pool of policies did not reach critical mass. Two separate EU commission reports have been released on the matter, from the Danish Patent Office and Trinity College Dublin, which produced contrasting models for the protection of patents for SMEs.[28] The UK Patent Office has also raised proposals for its own government-backed but subscription-funded mutual which would consider potential infringements to members' IP rights on their merits and purchase IP insurance where required to enable patent rights to be enforced.[29] The Danish study estimates that European welfare gains from a centralized patent scheme would be in the region of €6 billion to €21 billion.

The Danish Proposal argues for a European insurance policy administered by a central European agency that would act much like a broker with relationships to underwriters in the private sector. However, as with many European schemes, it is likely to experience practical difficulties in implementation. The variance in national insurance laws, often requiring the use of the local language, and issues of tax are two such hurdles. However, the real concern must be how the agency considers generating a large enough pool of policies, with an opt-in scheme likely to be avoided by companies with strong IP rights and so raising the liabilities of the pool. The low-cost approach of either applying a fixed premium per patent and a restricted policy cover or a full-blown cover rated on each patent's merits which costs more, has also yet to be agreed by the parties in discussion.

There has been a further development in the field of IP insurance, however, from the Lloyd's market. In the words of the annual Betterley Report, 'this is a very exciting product, since it protects against an exposure not otherwise insured – loss of a key income-generating asset of a company'.[30] Such a first-party policy allows the company to protect the commercial value of its IP rights in the same way as it would insure its headquarters, plant or stock with a fire policy. The policy developed by the writer's employer, R. J. Kiln & Co. Ltd, provides IP value protection rather than legal expenses and liability cover. Should IP rights be lost as a result of a legal action, such as a patent invalidity suit, or become unexploitable due to certain legal or regulatory actions of government, the policy looks to provide for this loss of value. Legal actions brought by ex-employees might also be included.

This form of cover is fundamentally novel as it looks at IP from the perspective of a property right rather than a liability and requires a valuation of some kind being placed on the IP. Insured value can be derived through a portfolio valuation using accounting techniques, from using the 'premium' obtained by the IP in the form of sales or profit on a product integrating the IP, royalty and licensing streams, examining the R&D expenditure that created the IP or, indeed, a value assigned to the IP for a collateralized loan or acquisition. A sum insured or policy limit less than the total value of the IP might be of interest to an assured who wishes to fund a contingency plan following the loss of the IP. Depending on the risks, it may be worth insuring the entire IP portfolio of a company or only selected families of IP that have strategic value. A cover such as this provides bottom-line protection to IP assets which, increasingly, find themselves in company financial statements.

## Concluding remarks

Innovation brings considerable risk and reward to the company and although much of the value and risk to R&D expenditure occurs with the formation of intangible assets they have rarely, if ever, been treated in isolation by the risk management community or at board level. Not only has there been considerable external pressure for this to change but there is finally an increasing realization of the value of assets of an intangible nature. The risk management community must play a part in recognizing and protecting this value from the numerous potential risks, but ideally there will be a corporate strategy for intangibles that looks both to protect and exploit value and which is supported through the company by accountants, lawyers and those responsible for corporate governance.

The insurance industry, maybe uncharacteristically, has responded to the perceived need for cover rather than the actual demand which has been slower to develop than it should have been. However, work must still be done in the area with a particular effort made to encourage more carriers to enter the IP insurance market and so increase capacity. Only with further efforts by risk managers, insurers and brokers will the market be able to develop and support industry in its innovation progress.

## Notes

1  www.innovation.gov.uk.
2  'The 2004 R&D Scorecard', UK Department of Trade and Industry, October 2004.
3  Hadjiloucas and Winter, 'Reporting the Value of Acquired Intangible Assets', IP Value 2005, *IAM Magazine*, Globe White Page Ltd.
4  Interbrand, 'The Global Brand Scorecard 2004', www.interbrand.com.
5  Abraham Lincoln.
6  Nakamura, 'What Is the U.S. Gross Investment in Intangibles? (At Least) One Trillion Dollars a Year!', Working Paper no. 01-15. Federal Reserve Bank of Philadelphia, 2001.
7  Earl Wilson, US newspaper columnist, Field Newspaper Syndicate.
8  *History of Economic Analysis. Capitalism, Socialism and Democracy*, New York: Harper Torchbook Edition, 1976.

9   For a useful and brief trip through prevailing theories, see Lévêque and Mémière, 'The Economics of Patents and Copyrights', www.cerna.ensmp.fr/PrimerForFree.html.

10  Nordhaus, 'The Optimum Life of a Patent: Reply', 62 *American Economic Review* 428, 1972.

11  Sakakibara and Branstetter, 'So Stronger Patents Induce More Innovation? Evidence From the 1988 Japanese Patent law Reforms', 32 RAND *Journal of Economics* 77, 2001.

12  Implementation can be noted with the issuance of FAS 141 and 142, integrated in the UK under FRS 10 and 11, or SFAS 141 and 142 in the US.

13  www.knowledgeleader.com, The Internal Audit and Risk Management Community.

14  Swiss Re's Patent Insurance Database, cited in Swiss Re promotional material.

15  An excellent discussion of this history and position on 'employee inventions' can be found in Page, 'A Global Issue Played Out in Japan', p. 17, *Intellectual Asset Management Magazine*, Issue 9, 2005, Globe White Page Ltd.

16  Hiroshi Okuda, Chairman of Japan Business Federation, quoted in, 'Settlement in LED Lawsuit Could Spark Surge in Lawsuits by Inventors', www.asahi.com.

17  As note 15, p. 18.

18  Taken from Wild, 'Investors Learn to Appreciate the Value of IP', IP Value 2005, *IAM Magazine*, Globe White Page Ltd.

19  Brackner, 'Returning Buoyancy in Market Conditions', p. 19, IP Value 2005, *IAM Magazine*, Globe White Page Ltd.

20  'Howrey: A Survey of Investor Attitudes on IP Protection', 2002.

21  DLA IP Survey of European Companies, October 2004.

22  Accenture & Economist Intelligence Unit, 'Survey Results on Accounting for and Managing Intangible Assets', 2003.

23  'The Aon European Risk Management & Insurance Survey 2002–2003'.

24  Aon Ltd, 'Biennial Risk Management and Risk Financing Survey 2003', London.

25  Ibid.

26  For some of the best analysis consider, Lanjouw and Schankerman, 'Enforcement of Patent Rights in the United States', from Cohen and Merrill, *Patents in the Knowledge-Based Economy*, The National Academic Press, 2003.

27  IP Wales, 'Intellectual Property & Legal Expense Insurance', www.ipwales.com.

28  Danish Ministry of Trade and Industry, 'Economic Consequences of Legal Expenses Insurance for Patents', 2001; and European Commission, 'Enforcing Small Firm's Patent Rights', 2001, Directorate-General for Enterprise, EUR 17032.

29  See Michael Edwards & Associates, 'Scoping Study: Report of the Patent Enforcement Project Working Group', June 2004.

30  Betterley Risk Consultants, Inc., 'The Betterley Report: Intellectual Property Insurance Market Survey 2004', April 2004.

# 4 Developments in patent enforcement procedure in Japan and England

*Anthony Trenton*[1]

## Introduction

The patent system is one of the foundations of a knowledge-based economy. It is essential for countries with such economies to provide effective patent enforcement procedures. Ineffective procedures increase risk for technology-based companies, whether by failing adequately to protect companies' investments in their innovations (thereby undermining the very purpose of the patent system) or by sustaining wrongly granted patents that provide a potential barrier to technology companies seeking to enter the market or develop other technology.

The use of patents in encouraging technological innovation has been recognised since at least 1432 when the Statute of Venice was passed. The principle is simple. In exchange for disclosing its invention to the public, the patentee is rewarded with a monopoly for a certain period of time. As a monopoly is potentially harmful to the economy, it is awarded only if certain conditions set out by patent law are met – for example, the invention must be new and not obvious. The aims of the patent system are at least twofold. First, armed with the knowledge that if he succeeds he will be rewarded with a monopoly, the inventor is encouraged to develop new inventions. In particular, an inventor is encouraged to invest in development. If it were all too easy for competitors to adopt an invention immediately after it had become public, there would be little or no incentive to invest in developing inventions in the first place.[2] Second, the inventor is encouraged to disclose his invention to the world, rather than seek to preserve a commercial advantage by keeping it confidential.[3]

A patent may be exploited in various ways. As with other property, it may be assigned. Also, it may be licensed, permitting others to use the invention on payment of a royalty. Or, of course, it may be enforced in the courts – where the patentee may seek an order preventing an infringing party from using the invention and an order for damages and other remedies. However, irrespective of how a patentee chooses to exploit a patent, the value of a patent will always be dependent on it being able to be enforced, if necessary. Court proceedings for enforcing patents and dealing with other patent disputes are therefore fundamental to the patent system.

The ideal for patent proceedings (whether for infringement actions, validity challenges, or indeed proprietorship disputes) is to be consistently reliable, fast and accessible (i.e. inexpensive). If the system does not possess each of these qualities, one way or another the effectiveness of patent enforcement will be diminished. Ineffective patent enforcement proceedings increase risk for the parties and the public. Slow proceedings clearly allow infringers to remain on the market and to continue to damage the patentee's interests. Equally, slow proceedings may deter a party wrongly accused of infringing for longer, or may keep invalid patents on the register for longer, thus affecting the public. It is also necessary for the system to be reliable. The dangers of an unreliable system are clear, leading to the risk of unjust decisions being made.

Patent law is a complicated area of law and more often than not involves technical facts. It is therefore advantageous for specialist patent judges to be involved in deciding patent disputes in order to ensure high-quality and consistent decisions are produced. The complex nature of the proceedings also means that they have the potential to be lengthy and costly. The other pressure is therefore for them to be fast and as inexpensive as possible. Both the Japanese and English systems have been making strides to achieve these ideals, and developments are afoot in both jurisdictions. This chapter explores the more recent developments in both jurisdictions that have been adopted with a view to achieving fast, accessible and specialist procedures.

## Japan

Japan was one of the world's economic success stories up until the end of the 1980s. One reason for this is that it was, just as it continues to be, at the forefront of technological developments. However, another major factor in its success was the ability of its hugely efficient and disciplined workforce to produce competitive and high-quality products. Since the economic bubble burst in the early 1990s, Japan has been unable to restore its international competitiveness. This is partly due to the competition from other Asian countries which has emerged as those countries are able to manufacture products at lower cost in light of lower cost labour. It is clear that Japan's traditional manufacturing-based approach, which was responsible for so much success in the past, is now less effective and cannot be relied upon for success in the future. It is now necessary for Japan to adapt to become a knowledge-based economy.

The Japanese government has appreciated the need to make the transition to a knowledge-based economy and hence to carry out necessary reforms.[4] Among other strategies implemented by the government recently is an IP strategy with the aim of making Japan a 'nation built on the platform of intellectual property' (per Prime Minister Koizumi). Following a policy statement by the Prime Minister on 4 February 2002, the Strategic Council on Intellectual Property was convened in Japan to establish and advance a national strategy for IP. The Strategic Council subsequently issued an IP policy outline and the following year the Intellectual Property Strategy Headquarters was established to replace the Strategic Council. The IP Strategy Headquarters consists of the Prime Minister, the Chief Cabinet Secretary and several ministers and a variety of experts from academia and industry. In July 2003, shortly after it was established, it issued the IP Strategic Programme (revised in May 2004). Among other things (and apart from the developments in the IP court system which are further discussed in this chapter), the Strategic Programme seeks to make universities more IP focused, speed up patent application examinations, increase anti-counterfeiting measures, support SMEs by reducing patent fees and subsidising applications for foreign patents, and increase the number of IP professionals in Japan, namely lawyers (*Bengoshi*) and patent attorneys (*Benrishi*).

Clearly, therefore, IP has become a priority for Japanese policy makers at the highest level. Furthermore, Japanese industry, which is keen to protect itself from products produced by Asian competitors at lower cost, has shown signs of attaching more importance to IP rights and their enforcement. In light of this, both government and industry have been keen to have a specialist patents enforcement system. As stated, patent law is very specialist and patent proceedings involve technical factual aspects. There is therefore a real advantage in having specialist courts to consider patent issues. With the IP strategy being of such importance at government level, this has now been achieved.

The recent changes which have been made to Japan's patent procedure relate (1) to the patent court system to make it more specialist and (2) to the way in which invalidity can be raised in proceedings and a patent revoked. In this chapter, first, the way things were previously is considered.

### The way things were

#### Validity

While a patent may be granted by a patent office that is not to say it is certainly valid. It may not be new or inventive in light of prior art (or an argument) that has escaped the examiner. The claims may be too broad for the invention that has been disclosed, or the patent specification may otherwise not enable the invention to be performed. The invention may also not be industrially applicable. If a patent is found not to be valid, it cannot be infringed.

Previously, the validity of a Japanese patent could only be challenged in the Japan Patent Office (JPO). Indeed, the Court was not able to consider validity at all during infringement proceedings. If infringement proceedings were commenced, they would sometimes be stayed pending determination of the validity of the patent by the JPO. There were two procedures that could be carried out in the JPO – the opposition procedure and the invalidation appeal. On occasions, both procedures could be invoked against the patent, although the rules as to who was eligible to commence such proceedings and the time limits differed. Under the opposition procedure, it was open to anybody to oppose the validity of the patent within six months of the advertisement of the granted patent in the Patent Gazette. Under the invalidation procedure only an interested party was able to challenge the validity of the patent. There was no time limit for doing so under this procedure. Both procedures were conducted before a body of three or five examiners of the JPO. Under the opposition procedure a dissatisfied patentee could appeal the body's decision to the Tokyo High Court; the opponent, however, had no means to appeal. Under the invalidation procedure, either party could appeal the JPO's decision to the Tokyo High Court.

In 2000, the Supreme Court of Japan decided in *Fujitsu* v. *Texas Instruments* that validity grounds could be raised in infringement proceedings. If there were convincing grounds for considering the patent to be invalid, then the infringement action would fail. This was important as it allowed 'squeezes' to be raised, in which the patentee cannot argue on the one hand that the claim is broad enough to cover the alleged infringing article, if that breadth also means that it renders the patent invalid (for example, by covering prior art). However, the Court still did not have power formally to decide that the patent was invalid – merely to dismiss the infringement action on the basis that the patent appeared to be invalid.

*Infringement*

The first instance court for patent proceedings in Japan is the District Court. Appeals from the District Court can be made to the High Court. A final appeal, on questions of law only, rests with the Supreme Court of Japan.

Previously, each District Court (of the 50), no matter how little expertise in IP matters it had, had exclusive jurisdiction to hear patent disputes in its district. However, from 1998, Tokyo District Court was given concurrent jurisdiction with eastern District Courts for all eastern areas of Japan and Osaka District Court was given concurrent jurisdiction with the western District Courts for all western areas. All patent disputes could still be heard in the other District Courts, however the majority of all patent cases were, from that time, heard in Tokyo or Osaka. These two District Courts have specialised divisions dealing with IP (for example three in Tokyo) and the first tier courts therefore, in practice, became more specialised.

Appeals from any particular District Court lay to the regional High Court having jurisdiction over the district. This led to appellate courts with little expertise having jurisdiction over patent cases, however, again, after 1998 most appeals were heard in Tokyo and Osaka.

*Recent developments*

As stated, with the importance of IP being acknowledged by both government and industry, there has been pressure for the courts to become more specialised in dealing with, in particular, patent disputes. After much debate and discussion, consensus as to the way forward has been reached.

From April 2004 some major changes were made to the Civil Procedure law. First, all litigation relating to patents, utility models, circuit design rights and copyright in computer programs is to be assigned exclusively either to the Tokyo District Court or Osaka District Court. Accordingly, it will no longer be possible to bring proceedings in other District Courts which previously had concurrent jurisdiction with those two courts. The number of specialist divisions in the Tokyo District Court has been increased to four to meet the increased demand. In addition, all appeals will now be heard by the Tokyo High Court, which has by far the most specialisation and expertise in the IP field. The number of judges in the IP division of the Tokyo High Court was increased from 16 to 18 to meet the additional demands.

Second, in the Tokyo High Court, a grand panel system was introduced to ensure consistency of High Court decisions. When cases raising the same issues are pending in the High Court, they will be heard by a panel consisting of five leading judges with IP expertise.

Third, 140 technical advisers have been appointed to assist the High Court and District Courts on technical matters. The technical advisers are university professors or researchers in public or private organisations who will be able to assist the court with their expertise on a part-time basis when required. This is in addition to the full-time technical clerks who are already employed by the courts (who are former patent examiners or patent attorneys used to assist on technical matters).

Since April 2004 therefore, substantial practical changes have been made to the patents court system to make it more specialist and hence reliable. However, in 2005 an amendment to the law comes into effect creating a new IP High Court, as part of the Tokyo High Court. This, in fact, is something of a compromise as some had wanted an entirely independent IP High Court. In any event, as a practical matter, it is unlikely to make any difference from the situation as from April 2004. Nevertheless, the amendment to the law and the formal creation of the court has the effect of enshrining in law the new position with respect to IP specialisation, which formality is seen as appropriate in light of the importance being attached to IP rights at present in Japan.

Another major amendment made to Japanese law that comes into effect in 2005 is to allow invalidity to be raised formally as a defence to infringement. Rather than arguing merely that the patent appears to be invalid and thus should not be enforced, which has been the position since *Fujitsu* v. *Texas Instruments*, the Court will be able formally to decide that the patent is invalid. This decision however will be binding only on the parties as the power to revoke the patent will remain with the JPO.

In addition, the former position whereby there were two ways of challenging validity in the JPO is being abolished. There is now to be only one procedure, a revised invalidation procedure. Under this procedure, any person can challenge the validity of the patent, and there are no time limits for doing so. The decision of the JPO may be appealed to the IP High Court.

In summary, these important changes have made the Japanese patent system more specialised and better able to cope with this area of law with its complex facts. The changes to the procedures for attacking the validity of patents are also important as they render proceedings fairer and more efficient. No doubt there will be many more developments in the future, but these changes will help to ensure that Japan has a fast and reliable patent enforcement system.

## England

The UK has also recognised the importance of IP and the need for a reliable court system in which patent disputes can be litigated. The use of specialist courts and judges to deal with patent disputes has up to now been a longstanding feature of the English system.[5] Patent infringement cases at first instance have, for a long time, been heard by specialist judges in the Chancery Division[6] of the High Court and appeals from UK Patent Office decisions were, since the Patents and Designs Act 1932, heard by the Patents Appeal Tribunal, which consisted of a nominated specialist judge (who tended to be one of the specialist judges hearing patent proceedings in the Chancery Division).

In 1977 the Patents Court was constituted by the Patents Act of that year.[7] This court is part of the Chancery Division of the High Court and consists of (usually) five nominated judges. They have (save in exceptional circumstances) exclusive jurisdiction to hear all patent matters in England and Wales. The Patents Court hears patent infringement matters, as well as appeals from decisions of the UK Patent Office (previously heard by the Patents Appeal Tribunal which was abolished).[8]

Appeals from the Patents Court lie (with permission) to the Court of Appeal. The Court of Appeal will usually sit in divisions made up of three Lord Justices of Appeal, one of whom is usually a former Patents Court judge. This enables the Court of Appeal to have a degree of specialism. An appeal may lie to the House of Lords if a question of law of public importance arises (however, patent cases only rarely proceed to this second tier of appeal).

It is to be noted that in England the formal specialist court is at the first instance level. The position is to be contrasted with that in Japan whereby the formal IP Court is constituted at appeal level (although, in both jurisdictions, specialist judges will hear the case at both levels).[9] The reason for this is that English civil proceedings are particularly detailed at the first instance stage. They involve experiments, disclosure by each party to the other of documents, exchange of expert reports on witness statements, skeleton arguments setting out the arguments of each side, full cross-examination of expert witnesses and factual witnesses and full oral argument. In light of the specialist area of law and often complex technical facts that the judge will have to decide upon, it works well to have specialist judges at first instance (if necessary, assisted by a scientific adviser). The appeal to the Court of Appeal is by way of a review, rather than a rehearing, and the Court of Appeal will only rarely reverse the court below on a question of fact. Rather, it is on grounds of law (including the interpretation of the patent) that a judgment is more likely to be reversed. Accordingly, there is no specialist appeal court and a single specialist judge (sitting with two other judges) is sufficient. Nonetheless, a specialist appeal court, or at least having more than one specialist appellate judge, as is the case with the new Japanese system, would have advantages as it would allow for a greater degree of informed discussion (or even dissent) among appellate judges.

The need for a specialist patent court system has therefore, to date, been met in England. Although the proposal in the Department of Constitutional Affairs consultation paper is certain to be controversial as the UK debates the need for a statutory specialist court, there is no suggestion of allowing non-specialist judges to hear patent cases. With its specialist judges, disclosure, experiments, cross-examination and thorough oral proceedings, the jurisdiction is recognised as a reliable high-quality one in which to resolve a patent dispute and will undoubtedly remain so, whatever the outcome of the consultation paper. Such detailed litigation procedures, however, can potentially lead to slow proceedings and high costs. In striving to achieve a reliable fast and accessible system, the issues to be resolved have, thus, in the past, related more to speed and cost. While it is of course vital for decisions to be of high quality and to be consistent, as they have been, it is also important for parties and the public for disputes to be resolved cheaply and quickly. Various major developments have taken place over the past fifteen years with a view to speeding up and reducing the cost of patent proceedings, particularly so in recent years.

### Developments in the late 1980s and the 1990s

In 1988, the Patents County Court was set up.[10] While county courts are, together with the High Court, courts of first instance, they are aimed at lower value and less complex cases. The Patents County Court was therefore constituted with a view to providing an alternative forum to the Patents Court for lower value, usually simpler, cases. Unfortunately, it did not initially succeed in its purpose; however it has recently been revitalised and is now a real alternative to the Patents Court in appropriate cases. As regards procedure, however, since the Civil Procedure Rules were passed in 1999, county courts and the High Court have the same procedure and therefore there is now no discernable difference in this respect. Also, if the outcome of the consultation paper on establishing a single civil court referred to above is to merge the High Court and county courts, clearly the Patents County Court will cease to exist.

In 1995, the Patents Court made some major changes to the way in which discovery of documents (as it was then called) was to be given to the opposing party in patent litigation. The general rule at the time, applicable to all civil cases, was that any document that was in the possession, custody or power of the parties and which related 'to matters in question in the action' had to be disclosed.[11] This included any document that contained information that may enable the other party 'either to advance his own case or to damage that of his adversary, if it is a document which may fairly lead him to a train of enquiry which may have either of these two consequences'.[12] The reason for this is that English civil proceedings (and those of other common law jurisdictions) are carried out with 'cards on the table' and any relevant document should be brought to the attention of the other side. However, the scope of discovery was often vast in patent cases, even though many documents had only the remotest relevance. This invariably slowed down proceedings and added to the cost.

A change to the rules in 1995 meant that, on the issue of validity, only documents produced within a window defined by the dates two years either side of the patent in suit's priority date needed to be disclosed (i.e. four years' worth of documents). Thus, insofar as details of research projects are disclosed (some of which, for example, may need to be disclosed as they relate to the obviousness of the invention, or to the extent to which it is enabled), only four years' worth of documents need to be disclosed rather than the entire project. On the issue of infringement, the alleged infringer may serve a product or process description setting out full particulars of what he does, instead of giving every document relating to his product or process. These steps have resulted in the scope of disclosure being radically cut down in many cases; the scope of disclosure was thus vastly different from that in the US, while still being substantial enough to uncover the most relevant documents that may turn out to be crucial to the case. This has had the result of reducing costs without sacrificing what is considered to be an important part of the investigative process.

Also in the mid-1990s, the patents judges began to take more control over the procedural timetable of proceedings, rather than leaving it only to the parties to agree the various deadlines leading up to trial. Leisurely timetables (which one party may seek at the expense of another, or both parties may seek – at the expense of the public) are no longer tolerated and a trial date will be fixed at the very early stages of the case. More often than not a trial will take place within a year of the proceedings being commenced, often less. Appeals to the Court of Appeal will then be heard usually within a year of judgment. With proceedings at first instance taking place within a year, rather than a few years as was the case previously, a fast high-quality result can now be obtained from the Patents Court. Moreover, as proceedings are dealt with in a shorter time period, they tend to be more focused and hence the cost is lower.

Following Lord Justice Woolf's (as he then was) review of civil procedure in the 1990s,[13] civil procedure generally was radically changed in April 1999 with the coming into force of the Civil Procedure Rules 1998. A wide variety of changes were introduced. Among them was a change from the '*Peruvian Guano*' discovery referred to in note 12, to what is called 'standard disclosure'. This limits disclosure of documents to those that are to be relied on by a party or that adversely affect either party's case or support the opposing party's case. The notion of disclosing documents that may merely lead the opposing party on a 'train of enquiry' was thus ended. This has further reduced the scope and cost of disclosure without prejudicing its essential purpose. Another major change brought about by the Civil Procedure Rules is that the Court is to take a far greater role in case management including managing the procedural stages of the case. However, as the Patents Court, being relatively independent, had already been doing this for some years as referred to above, little change was discernible as a result of this general rule change.

### Recent developments

As a consequence of the changes that took place in the mid- to late 1990s, patent proceedings in England were far faster and were less costly. More recently, there have been further developments with a view to making patent proceedings in England even less costly and hence more accessible to SMEs and also to tailoring the procedure to make it more proportionate to the type of case being considered.

In April 2003, the Patents Court introduced a 'streamlined procedure'. The streamlined procedure is one in which all factual and expert evidence is in writing (that is to say there is no cross-examination, save on any topic where it is necessary), there is no requirement to give disclosure of documents, there are no experiments, the date of the trial is fixed to be normally within six months of the time when the streamlining order is made and the total duration of the trial is fixed, normally for not more than a day. Variations of the procedure are possible. The aim of this procedure is to make available very inexpensive proceedings, which may be appropriate where the more searching aspects of the usual procedure are not necessary. The Court will order the procedure by agreement of the parties or where it decides that it is just to do so bearing in mind proportionality, the financial position of each of the parties, the degree of complexity of the case and the importance of the case. There is also a duty on the parties' legal advisers to draw their clients' attention to this new procedure.

While this procedure strips away many of the more costly elements of the traditional procedure and thus makes litigation more accessible to parties in a weaker financial position and more proportionate where there is less at stake, it does so, of course, at the expense of the more searching aspects of the enquiry. Nevertheless, it does allow for a great deal of flexibility, enabling the procedure to be tailored according to the type of case. It is fair to say that parties in England do not appear to have taken up this procedure much so far. This may be due to conservatism and it is still early days. The option is there if desired and it remains to be seen if things will change and if it becomes more popular in the future.

The most recent development as at the time of writing is that in July 2004 the Patents Act 2004 was passed by Parliament, amending various provisions of the Patents Act 1977. The new Act provides for a new procedure whereby non-binding opinions from the Patent Office as to the novelty, obviousness or infringement of a patent may be requested by any party (including the patentee).[14] It is also possible for the patentee (or exclusive licensee) to request a review of such an opinion by the Comptroller (head of the Patent Office). The aim is for the procedure to be fast and simple. It has to be said that it is unlikely that any substantial patent dispute between parties in a strong financial position will be resolved by this non-binding procedure. However, in cases where both parties would prefer to settle disputes without having full proceedings, this may offer an alternative way forward.

Finally, the Patents Act 2004 has amended section 106 of the 1977 Act to provide that the financial position of the parties must be taken into account by the Court when awarding costs in infringement and validity proceedings. Ordinarily, the general rule in litigation in England is that the loser of the case will pay the winner's costs, insofar as they are reasonable and proportionate. This amendment raises the possibility that a losing party that is in a weak or relatively weak financial position may not have to pay all of the costs of the winning party. It remains to be seen how the court will apply this in practice, but clearly it is a further way of enabling SMEs to have better access to the Patents Court, where previously they may have been discouraged by the potential cost.

As regards the future, potentially the most drastic change for the UK is the proposed Community Patent or, alternatively, the separate European Patent Litigation Agreement (EPLA). The former proposal is for an EU-wide single patent (applied for through the existing European Patent Office in Munich) the validity and infringement of which will be decided in a Community Patent Court in Luxembourg which will have its own procedure (with an appeal to the EU Court of First Instance, also in Luxembourg). This proposal is aimed at cutting down on the expense of enforcing patents in multiple jurisdictions in Europe and the uncertainty and inconsistency that can result from multinational litigation. As at the time of writing, however, the negotiations over the Community Patent appear (once again) to be faltering over the issue of translations.[15]

If the Community Patent proposal does falter, the EPLA may well proceed.[16] This relates to the existing system of European patents which are granted by the European Patent Office (EPO) under the European Patent Convention (EPC). Upon grant by the EPO, the European patent becomes a bundle of national patents which, subject to post-grant opposition at the EPO where they can all be revoked, have an independent life from each other. As things stand, subject to the opposition procedure, it is necessary to apply to revoke the patent or sue for infringement of the patent in each relevant designated country. The signatories to the EPLA would agree to the creation of a unified judicial system for dealing with litigation relating to European patents. It would involve the establishment of a European Patent Court with a central chamber (the site of which is undecided) and regional chambers, which would have the power to revoke European patents across all designated states which are signatories to the agreement and to decide on infringement. Appeals would lie to a second instance European Patents Court. Panels of the first instance court would be made up of an odd number of judges and would be comprised of at least one technically qualified judge and at least two legally qualified judges, who will be of at least two different nationalities. If the EPLA does come to fruition, it will be a major change for England as it will include an approach to litigation more akin to that of Continental European systems.

## Conclusion

Both Japan and the UK clearly recognise the importance of IP and hence the corresponding importance of having a court system that is reliable, consistent, fast and accessible. These qualities are important for the parties and important for the public. In both countries there have been recent developments in the patents court systems that lead to this aim. Japan has, among other things, focused on making the court system more specialist, as well as simplifying the procedures for making invalidity challenges. The English courts have long been specialist (although, as stated, there is to be a consultation on whether the statutory basis for the Patents Court should remain). Over the years, the courts in England have been concentrating on making their procedures more proportionate where appropriate, and faster. The most recent developments of 2004 in England emphasise even more the importance that is attached to enabling parties in a weak financial position to have access to the court. Undoubtedly, both jurisdictions' courts' procedures will continue to develop to enable parties and the public to obtain decisions reliably, quickly and inexpensively and other factors may also cause major changes in the future, such as those relating to a Europe-wide system.

## Notes

1   Anthony Trenton has been assisted in writing the present chapter by Mr Yasukazu Irino, currently Commercial Counsellor at the Embassy of Japan in London. Mr Irino was formerly the Deputy Director of the Policy Planning Unit at the Japan Patent Office. Acknowledgements are also due to Judge Ryuichi Shitara of the Tokyo High Court for his kind permission to use material presented in 2004 at the 13th Annual Fordham University International Intellectual Property Policy and Law Conference which was organised by Professor Hugh Hansen, Professor at Law, who has also given his permission for this material to be used.

2   Abraham Lincoln put it this way during a lecture on 'Discoveries and Inventions' given on several occasions in Illinois in 1858 and 1859:

> Before [the patent laws], any man might instantly use what another had invented; so that the inventor had no special advantage from his own invention. The patent system changed this; secured to the inventor, for a limited time, the exclusive use of his invention; and thereby added the fuel of interest to the fire of genius, in the discovery and production of new and useful things.

3   In fact, the European Commission is carrying out a study to consider the effect that patents have on the EU economy and society.

4   This topic is further explored in the book, *Exploiting Patent Rights and a New Climate for Innovation in Japan*, ed. Ruth Taplin (London: IPI 2003).

5   The procedural system in Scotland is not considered in this chapter.

6   This division deals with issues concerning the law of land, probate and trusts, etc. The other two divisions of the High Court are currently the Queen's Bench Division (general civil work, but also contains the Commercial Court and Admiralty Court) and the Family Division.

7   By section 96 of the Patents Act 1977, which was subsequently repealed and replaced by section 6 of the Supreme Court Act 1981, which also forms the Commercial Court and Admiralty Court.

8   However, it should be noted at the time of writing the UK Government's Department of Constitutional Affairs has produced a consultation paper proposing the merging of the High Court and county courts to form a single court ('A Single Civil Court? The scope for unifying the civil jurisdictions of the High Court, the county courts and the family proceedings courts', 3 February 2005). One of the proposals is to remove the statutory basis of the Patents Court. The paper proposes retaining a patents list and specialist judges. This author suggests that this would be a retrograde step. The paper proposes retaining the Commercial Court as it has international prestige. It is submitted that the same reasoning should apply to the Patents Court. At the time of writing, it is not clear what support the proposal will have.

9   The US is different still. First instance cases are heard (often as jury trials) in various District Courts, which are not specialist, but some of which have more experience in patent matters than others. All appeals are then heard by the Court of Appeals for the Federal Circuit (CAFC) in Washington DC which was set up to hear appeals in certain specialist areas of law, including patents.

10   Under section 287 of the Copyright, Designs and Patents Act 1988.

11   Rules of the Supreme Court Order 24, Rule 1.

12   *Compagnie Financière du Pacifique* v. *Peruvian Guano Co.* (1882) 11 QBD 55.

13   The Access to Justice report.

14   Section 10 of the 2004 Act which inserts sections 74A and 74B into the 1977 Act.

15 Translations have been an ongoing issue in negotiations relating to the Community Patent. In 2003 it was finally agreed by Member States that only the claims need to be translated into each language of the EU, rather than the entire specification. However, there have since been issues raised relating to the timing for translating the claims and also whether the translated claims or the original claims are to be definitive.

16 It too has stalled – pending the negotiations for the Community Patent due to opposition from the EU Commission and some Member States.

# 5 Risk transfer in a changing world

*Oliver Prior*

In the wake of the hardening insurance market, precipitated by the events of 9/11, many traditional insurance buyers find themselves being faced with having to self-insure significant levels of risk or, alternatively, they are electing to retain more risk because their own 'cost of risk' valuations are less than the equivalent imputed by insurers when calculating insurance premiums. At the same time a new breed of potential insurance buyers/drivers has emerged, namely financiers and regulators. The next few years will be critical for the insurance industry since it will need to re-appraise its offerings to traditional insurance buyers to reflect the modern risk managed environment and, at the same time, to radically alter some longstanding insurance traditions if it is to meet the needs of financiers and regulators who seek certainty and 'pay now argue later' risk transfer contracts. This chapter considers how traditional insurance buyers' attitudes to risk transfer have changed, or been changed, in recent years and how regulators and rating agencies have come to recognize the existence of 'operational risks' and are creating an environment where risk transfer alternatives are increasingly being sought.

## Risk segmentation

With accurate statistical data available for analysis most insurable 'losses' divide into three categories (Figure 5.1):

Level 1 Losses that can be absorbed within the annual budget of a business or business unit's profit and loss accounts.

Level 2 Losses that occur every three to ten years, with reasonable predictability, but cannot be managed within the annual budget of a business's annual profit and loss account.

Level 3 Risks where losses rarely occur, however when they do the impact on the business's balance sheet can be catastrophic.

To add another dimension such 'losses' need to be divided into two categories, namely:

- damage to assets ('short tail'); and
- potential and/or actual corporate liabilities ('long tail').

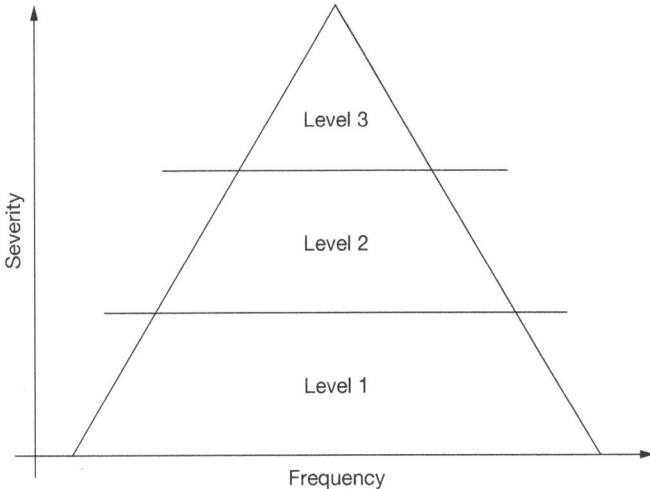

*Figure 5.1* Risk segmentation

In the case of a short tail loss an estimate of actual loss can usually be achieved within a fiscal period, an appropriate specific 'reserve' can be established and, in some instances, indemnification can be achieved. With a long tail loss such speed and accuracy are rarely achieved within a fiscal period. Quantum and potential liability can only be estimated and a reserve established today and the actual loss only becomes known in a future fiscal period. Thus, in the case of corporate liabilities reliance on current data often involves working with specific 'reserved' figures yet using only settled or part settled claims involves using out-of-date frequency factors. The key to useful data is the 'triangulation' of historic long tail claims performance in such a way that the triangulation will illustrate the average gestation period of certain liability allegations and show how many allegations translate into actual settlements. These data can then be applied to current reserves in such a manner that an expected maturity pattern will emerge. Quarterly review of reserve estimates throughout the life of such allegations can also help to illustrate average trends.

Figure 5.2 illustrates the typical 'hockey stick' curve that results from a normal long tail triangulation analysis.

Statistical data are the key to a reasonably accurate determination of the beginning and ending points of Levels 1, 2 and 3.

We can assume the following:

Level 1  will be 'self-insured'/retained by the business or business units;
Level 2  a mechanism is required that will cost-effectively spread the business's predictable, but not annual, losses equally across a series of fiscal periods and facilitate the effective establishment of non-specific reserves;

Level 3   a mechanism is required that will 'pool' the risk of unexpected loss among similar like-minded organizations in exchange for a premium payment.

Companies will often make use of a Captive Insurance Company, Protected Cell Captive Insurance Company and/or Alternative Insurance option for the area of risk in Level 2, whereas the conventional insurance industry is usually called upon to provide protection against loss resulting from risks in Level 3.

In practice however many insurance buyers are no longer as dependent upon the offerings of the conventional insurance industry as they used to be for three reasons:

1   for certain types or organization the capacity of the insurance industry is often insufficient to manage a truly catastrophic risk; and
2   the insurance industry has developed an aversion for 'systemic' risk events; and
3   the 'banking' industry is offering increasingly sophisticated contingent capital/finance alternatives.[1]

Many organizations, such as financial services and energy companies, do not believe that the current market capacity for catastrophic risk events is sufficient for their needs. Catastrophes in such industries are viewed in terms of billions of dollars of potential loss and in many cases the insurance market's capacity struggles to reach one billion dollars; particularly in a hard market. In some areas of risk, such as Y2K, computer virus, financial advice etc., where the insurance industry has identified that a

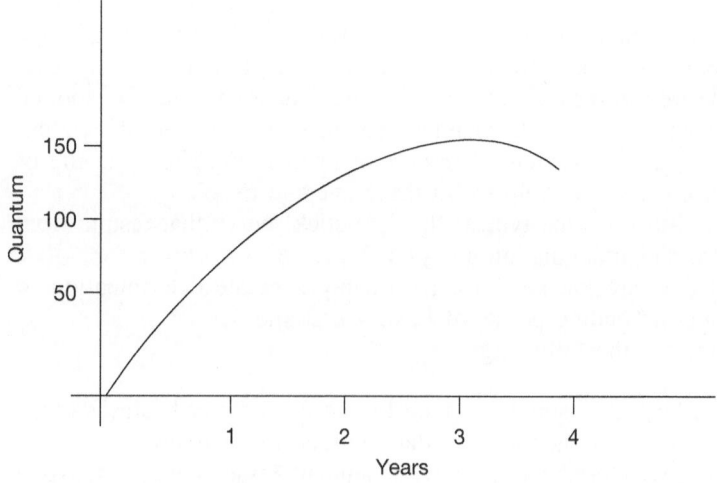

*Figure 5.2* Illustration of long tail triangulation

single event can affect a wide range of insurance buyers ('systemic risk'), insurers have given up trying to offer any meaningful protection. If this trend continues then the insurance industry will only remain relevant to mid-sized organizations and then only in the case of risks not of a systemic nature.

The insurance industry that previously owned the 'risk franchise' is in danger of becoming one actor in this increasingly complex production. What is being overlooked, or ignored, by the industry is that just because a risk cannot be insured it does not go away.

Corporate governance is the new driver of risk management and it is self-evident to most boards of directors that if, for example, the legal consequences of providing financial advice, the impact of a computer virus or similar so called systemic risk can no longer be transferred to a third party, an alternative method of managing the risk must be found – it cannot simply be ignored. As companies identify more and more areas of risk via good corporate governance techniques and analyse these using risk management processes, so, inevitably, they must determine how to manage the fiscal consequences of them. Where a risk cannot be transferred by means of insurance the choice is simple; it must be retained and thus the fiscal management of that item becomes an issue of risk financing, but what of the risks that can be transferred?

## Arbitrage buying

In the case of risks that can be transferred but a company elects to retain them, albeit by means of a Captive Insurance Company, Protected Cell Captive Insurance Company or an Alternative Insurance technique, the company needs to be assured that the choice is an informed one. Inevitably, assuming risk, with or without the support of one of the former techniques, will result in the company's shareholders being confronted with additional risk and it is therefore important that the board of directors, acting on behalf of the shareholders, ensures that the risks in question are analysed and form a definite view concerning the 'cost of risk'. An increasing number of companies therefore find themselves valuing areas of insurable risk and this, in turn, puts them in a position to become 'arbitrage buyers'.

What then is an arbitrage buyer? The answer is simple, namely, it is an organization that:

1  has the ability to assume risk, albeit with the support of a Captive Insurance Company, Protected Cell Captive Insurance Company and/ or Alternative Insurance option; and
2  has evaluated a risk or suite of risks and determined that the insurance industry, at a particular point in time, is charging a premium that is below the 'cost of risk'[2] applicable to those risks.

In other words the arbitrage buyer does not need insurance to survive but will buy it when the price of the product is lower than it own perception of risk.

The problem for the insurance industry is that the increasing number of arbitrage buyers is at the sophisticated end of the risk management scale, e.g. they are aware of the risks they face, their own risk profile and in many cases their perceived 'cost of risk'. Since insurance is always a business of averages, to lose such buyers to self-insurance techniques could lead to problems if this phenomenon is not addressed. The arbitrage buyer is usually a student of the insurance cycle and will be prepared for a change. Figure 5.3 illustrates the potential for arbitrage.

To be an arbitrage buyer of risk transfer products, offered by the insurance industry, a buyer needs to be prepared for the arbitrage opportunity. Hard markets can come and go in a relatively short period of time so taking full advantage of the change needs preparation.

There should be no misunderstanding – the arbitrage buyer is choosing to assume risk as opposed to transferring it because they firmly believe, after analysis, that the cost of transferring the risk is not a reasonable bargain. The assumption of risk with no fiscal management is a luxury that very few entities can afford. Most businesses will have shareholders, partners, lenders, bondholders etc. whose considerations the board of directors has a duty to take into account. In other words an increasing number of organizations have the ability to retain greater levels of risk but they need to employ techniques to establish non-specific reserves where this can be achieved.

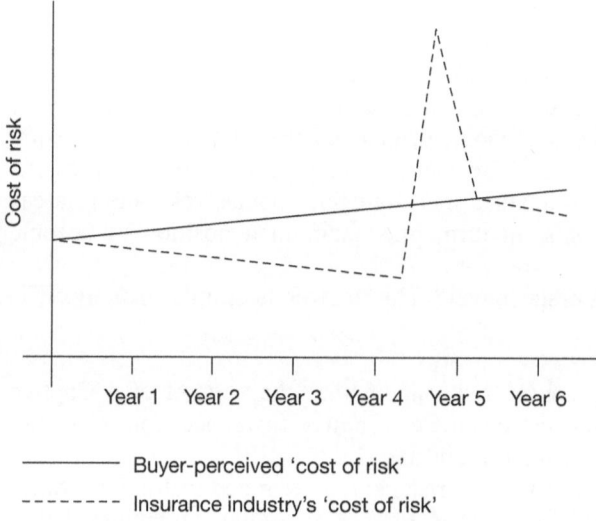

*Figure 5.3* Cost of risk vs. insurance cycles

A shareholder in a company that has chosen to assume more insurable risk, because it believes as a result of its own 'cost of risk' modelling that transferring a risk to the insurance industry is not in the best interest of the shareholders, should have nothing to complain about if they are aware of this decision (and the rationale behind it) and the board of directors has put in place mechanisms to ensure the fiscal impact of an event can be mitigated. The arbitrage buyer does not have to be 'loss free' to be right, since shareholders expect an element of risk when they purchase an equity interest, but the shareholders are entitled to expect good 'cost of risk' assumptions have been made prior to any decision to assume otherwise insurable risk and they are entitled to see that appropriate mitigation of loss techniques are being employed.

## Imposed self-insurance

Insurers are now seeking to exclude some previously insured perils or restrict cover on offer in an unacceptable manner. This means that the insurance buyers must now consider how they can best manage these increased levels of self-insurance. Businesses that previously turned to the insurance industry for all of their risk transfer needs now find themselves having to put in place the machinery of risk retention, e.g. Captive Insurance Companies, Protected Cell Captive Insurance Companies, Mutual Insurance Companies, Alternative Insurance structures etc., and in some notable cases are finding that their insurance operations are achieving underwriting profits when the insurance industry is failing to do so. Companies that used to rely heavily upon insurance find themselves learning to live with more risk assumption on their balance sheets.

## Managing self-insured risks

The most effective method of managing self-insured risk is still the use of a Captive Insurance Company. The term 'Captive Insurance Company' as used herein is a broad one and encompasses a number of different techniques such as Protected Cell Captive Insurance Companies, Mutual Insurance Companies, Risk Retention Groups etc. but they all have one thing in common, namely, they are vehicles established to 'warehouse' insurable[3] risk.

Commercial insurance companies rely on the basic principle of insurance, i.e. 'they seek to spread the losses of the few among the many'. Captive Insurance Companies rarely have the luxury of benefiting from sufficient spread of risk to be able to employ the basic principle of insurance so, instead, they must operate as a vehicle that spreads the impact of their insureds' loss(es) across several fiscal periods – a technique that is known as 'risk financing'. Commercial insurers have the ability to underwrite hundreds if not thousands of similar risks in the knowledge that it is unlikely that more than one or two of them will suffer a loss. When confronted with potential evidence of a possible systemic aggregation of loss such as Y2K or computer virus many of them 'walk away'. Captive Insurance Companies are therefore usually being asked to go where commercial insurers fear to tread. Clearly in such circumstances to copy the techniques of commercial insurance as regards pricing or reinsurance buying is not advisable.

By definition a Captive Insurance Company will usually be assuming greater 'risk' than a commercial insurer and in order to support that assumption of risk three basic tools are employed, namely, capital, premium and reinsurance. In the 1970s and 1980s many Captive Insurance Companies retained a minimal amount of risk and quota share reinsurance would usually be purchased – during these decades much stress was laid on the benefits of accessing the reinsurance market via a Captive Insurance Company. The primary source of income to the Captive was thus over-riders. Most Captive Insurance Company managers now realize this was a dangerous strategy since there have been many high-profile claim disputes between Captive Insurance Companies and their reinsurers and some notable (re)insurance failures. Today in the commercial insurance world 'fronting' costs 10 to 15 per cent of the premium income being fronted for and many 'fronters' will require 'AA' rated security and contractual waivers in the reinsurance contracts. At the same time many insurance buyers realize there is no such thing as a free lunch and the costs inherent in just using their Captive Insurance Company for 'fronting' will ultimately be passed on to them. If quota share reinsurance can be found then surely the provider of such reinsurance can be persuaded to deliver their capacity directly to the ultimate insurance buyer at more economic terms. The Captive Insurance Company is now, more often than not, used to retain risk and provide and/or access alternative reinsurance (colloquially referred to as 'finite').

The main difference between conventional insurance and the function of a Captive Insurance Company can best be described as

| Conventional (re)insurer | = | spreads the losses of the few among the many |
| Captive (re)insurer | = | spreads a business's own losses across a series of fiscal periods |

Figures 5.4 and 5.5 show this in diagrammatic terms.

Figure 5.4 shows how a conventional (re)insurer operates, e.g. in Year 1 Company 'B' faces a loss and Companies 'A', 'C', 'D' and 'E' do not so the premiums paid by all five companies finance the loss sustained by 'B'. In Year 2 Company 'D' faces a loss and Companies 'A', 'B', 'C' and 'E' do not. In each year premiums and losses should be equal. Figure 5.5 shows

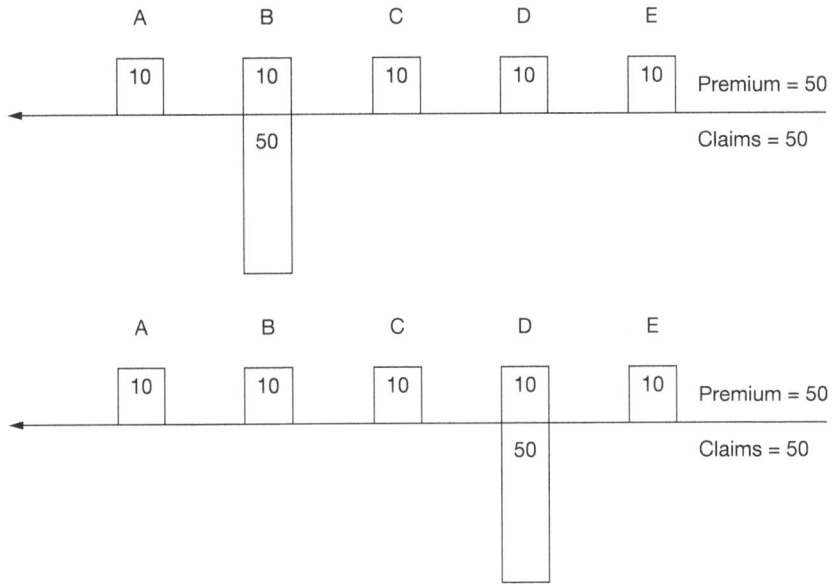

*Figure 5.4* Annualized premium and loss account examples

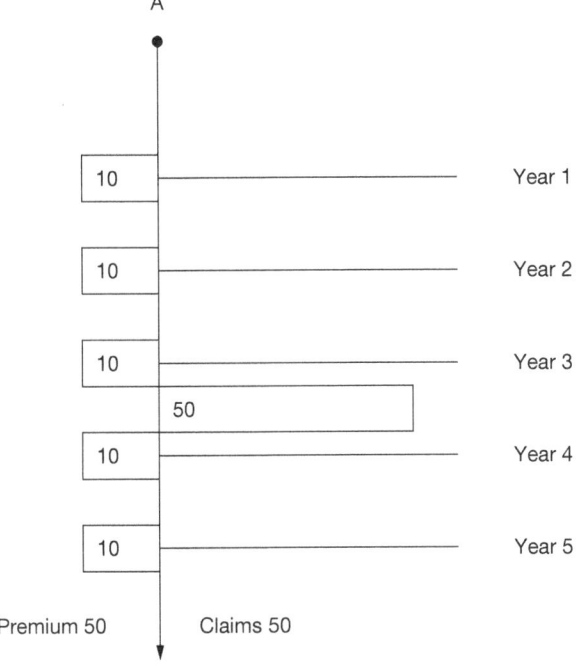

*Figure 5.5* Multi-year premium and loss account examples

how a captive (re)insurer operates, namely when Company 'A' faces a loss in Year 3 it has available a means to spread it across Years 1, 2, 3, 4 and 5 and so on to avoid the full impact of the loss on its balance sheet in the Year 3 account. Using this technique, non-specific reserves can be established to address unexpected mid-sized loss events.

It should be understood that conventional risk transfer (re)insurance and captive (re)insurance techniques can co-exist in one insurance buyer's portfolio; they are simply different methods of managing the fiscal consequences of 'risk'.

A captive insurance technique primarily transforms an insured peril into a credit risk. A captive (re)insurance contract provides a facility whereby the (re)insurer spreads the financial impact of a loss across a period of time. Captive insurance contracts are therefore principally about timing and credit risk as opposed to the primary cause of loss.

## Emerging risk transfer opportunities

Another issue facing the insurance industry in the modern world is that there are many new requirements for risk transfer mechanisms from financiers, rating agencies and regulators in support of 'operational risks' associated with the practice of securitization, project finance, PFI/PPP etc. where the transfer of risk is part of the rationale for the structure. The problem faced by such potential insurance buyers is the uncertainty of insurance. Insurance was founded in a different era and the principles of insurance that have stood the industry in good stead for centuries don't always work to its advantage today. By its nature an insurance contract is a conditional agreement between two parties that is governed by three main principles that are enshrined in law, namely those of 'insurable interest', 'indemnity' and '*uberrima fides*'. The principles of insurable interest and indemnity require that a party suffering loss shall have an interest in the loss and shall be no better off after an insurance claim payment than they were before it. Contrast this with a derivative where payment is made upon the happening of a trigger event irrespective of whether the purchaser has suffered any loss. Indemnity means that in the case of a loss the claimant is required to prove the amount of the loss suffered and this takes time – thus 'instant' settlement in the case of an insurance contract is virtually impossible. The principal of *uberrima fides* requires that the purchaser of an insurance contract must disclose all material facts to the insurer at the time of purchasing an insurance policy. The principle was designed to protect insurers from the need for them to conduct extensive due diligence reviews of every risk that they were offered. For most insurance contracts this principle works well, however where 'certainty' of payment is being sought it can lead to concerns by the party being asked to rely upon an insurance contract.

Following on from Barings and other notable bank problems of the 1980s and 1990s increasingly regulators are looking at 'operational risk' when considering risk-based capital models and this has recently become very evident in the consultation period for, and final publication of, 'Basel II'. In September 1998 the Risk Management Sub-group of the Basel Committee on Banking Supervision published a paper on Operational Risk Management[4] in which they sought to raise the awareness of operational risk as a topic. This first paper was a precursor to the publication of a consultative paper entitled A New Capital Adequacy Framework by the Basel Committee for Banking Supervision. This latter paper sought for the first time to define the scope of 'operational risk' and distinguish it from 'market risk' and 'credit risk', which were the main subjects of the first Basel Accord published in 1988.

The New Capital Adequacy Framework[5] (which was quickly labelled 'Basel II') explained that the final Accord, when published, would be seeking to bring 'operational risk' into the risk-based capital arena and while capital requirements overall may not be changed the total minimum capital requirements established under the First Pillar would, in future, be distributed across three areas of risk, namely:

- credit risk;
- market risk;
- operational risk.

Initially it was proposed that 20 per cent of a bank's minimum regulatory capital should be allocated to 'operational risk' but following consultation this was reduced to a target of 12 per cent.

In June 2004 the Basel Committee on Banking Supervision published the final version of 'Basel II', entitled 'International Convergence of Capital Measurement and Capital Standards',[6] and this document held very few surprises for those who had been following the consultative process.

Part II of 'Basel II' deals with the First Pillar, i.e. the total minimum capital requirements as they relate to credit risk, market risk and operational risk, and this section of the chapter specifically focuses upon Part V, namely, the provisions relating to operational risk and, in particular, how these recognize insurance as a mitigant to the minimum capital requirement.

'Operational risk' is defined as: 'The risk of loss resulting from inadequate or failed internal processes, people and systems from external events' – this definition includes legal risk[7] but does not include strategic risk or 'reputational risk'.

'Basel II' establishes three methods of increasing sophistication for calculating the minimum regulatory capital requirement relating to operational risk. These three methods are:

- The Basic Indicator Approach.
- The Standardized Approach.
- The Advanced Measurement Approach (AMA).

Internationally active banks and banks with significant operational risk exposures are expected to use an approach that is more sophisticated than the Basic Indicator Approach.

Throughout the consultation process experts in the field of operational risk management and financial institution insurance gave direct evidence and submitted written responses to the Basel Committee for Banking Supervision relating to the relationship between capital and insurance as it relates to operational risk.

In the event the Committee has acknowledged in 'Basel II' that it will allow AMA banks only to factor into their capital assessment models up to a 20 per cent reduction in their operational risk capital allocation provided that they can demonstrate that they have in force an operational risk insurance programme that meets the following key criteria:

- The insurance provider must have a minimum claims paying ability rating of 'A' (or equivalent).
- The insurance policy(ies) must have an initial term of no less than one year. For policies with residual terms of less than one year, the bank must make appropriate haircuts reflecting the declining residual term of the policy, up to 100 per cent haircut for policies with a residual term of 90 days or less.
- The insurance policy(ies) must have a notice period for cancellation of 90 days or longer.
- The insurance policy(ies) must not have any exclusions or limitations triggered by supervisory actions, or, in the case of a failed bank, that preclude the bank, receiver or liquidator from recovering for damages suffered or expenses incurred by the bank, except in respect of events occurring after the initiation of receivership or liquidation proceedings in respect of the bank, provided that the insurance policy(ies) may exclude any fines, penalty, or punitive damages resulting from supervisory actions.
- The risk mitigation calculations must reflect the bank's insurance coverage in a manner that is transparent in its relationship to, and consistent with, the actual likelihood and impact of loss used in the bank's overall determination of its operational risk capital.
- The insurance must be provided by a third party entity (in the case of a Captive Insurance Company or an 'in house' insurer being involved it must be shown that the risk has been reinsured to an independent reinsurer with a minimum claims paying ability of 'A' (or equivalent)).
- The framework for recognizing insurance must be well reasoned and documented.
- The bank must disclose a description of its use of insurance for the purposes of mitigating operational risk.

In response to the final consultation paper (known as CP3)[8] a number of organizations commented that restricting capital relief to operational risk transfer via insurance had certain disadvantages, namely:

- Alternative risk transfer techniques such as derivatives and insurance risk-based securitization, both currently in an embryonic state, were not being recognized.
- The Committee expressed its hope that a new generation of operational risk transfer contracts would be developed that more closely matched up to the needs of regulators and could play a greater future role in mitigating capital but by limiting relief solely to insurance-based contracts many think that developing derivative and insurance-based securitization products will not happen.

Despite these representations 'Basel II' continues only to recognize insurance as a mitigant but in a footnote it is stated that 'the Committee intends to continue an ongoing dialogue with the industry on the use of risk mitigants for operational risk and, in due course, may consider revising the criteria for and limits on the recognition of operational risk mitigants on the basis of growing experience' (the absence of the use of the term 'insurance' in this footnote could imply wider mitigation techniques being considered in future).

'Basel II', as it promised to do from the outset, has moved the subject of operational risk in banking from the corporate governance arena firmly into the area of 'risk-based' capital. At the beginning of the process most operational risk experts opined that the subject was far too subjective to lend itself to some of the quantitative disciplines that are employed to measure credit and market risk in banking. In the event 'Basel II' makes a good first attempt to introduce the subject of operational risk and it makes it clear to banks that adopt even the Standardized Approach that the minimum qualifying criteria are:

- its board of directors and senior management must be actively involved in the oversight of the operational risk management framework;
- it must have an operational risk management system that is conceptually sound and is implemented with integrity;
- it must have sufficient resources in the major business lines as well as the control and audit areas.

There is no doubt that the capture of operational risk within the new capital adequacy framework will open the subject up to extensive debate and, while some of the provisions of 'Basel II' with regard to operational risk may still be subjective, from past experience of how credit and market risk evolved after the publication of 'Basel I' it is clear that operational risk as a topic is here to stay. The door has been opened up for the risk transfer counter-parties such as insurers to develop a new generation of insurance contracts that will make the regulators more comfortable with the certainty of their products. The prize will be a greater share of the operational risk transfer market.

Insurance regulators, spurred on by the work of the banking regulators on 'Basel II', are now seeking to bring operational risk into the risk-based capital arena for insurance companies. The FSA in the UK have stated that insurance regulation is likely to move in the direction of 'Basel II', i.e. risk-based capital that seeks to address all areas of risk inherent in Insurers. This approach will favour insurance companies that know their risks and are able to model them in a manner similar to that proposed in 'Basel II' for banks.

Securitization proponents face a different set of circumstances but the roots of the problem that they confronted were to be found in the same principles of insurance. In any securitization process, where an investment grade bond issue is the required end product, the sponsor will need to address the various risks identified by the rating agency(ies) during the rating process. In the Hollywood Funding securitization a major problem was identified. The Hollywood Funding issue related to the securitization of a slate of film finance transactions that were supported by an insurance contract (known in the insurance industry as 'bums on seats' insurance). The insurance purported to indemnify the bondholders should the film-makers not recover sufficient monies from the box office receipts etc. to repay the loans – hence the euphemistic name given to such insurance by the insurance community. The insurance was issued by an investment grade rated multi-line insurer and on the back of the insurance in question the bond issue was imputed the same rating as that of the insurer. When a loss occurred, however, the insurer disputed payment under the insurance contract, citing *uberrima fides* as the primary reason for seeking to void the insurance contract (voidance of the contract of insurance is the only legal remedy available for a breach of the duty of *uberrima fides* on the part of an insurance buyer). This led to a very dramatic down-grading of the bonds and much analysis of the failings of insurance as a support mechanism for financial transactions. Despite the problems high-lighted by this latter example many financiers have little or no choice but to rely upon insurance contracts to transfer elements of risk associated with overseas lending, project finance etc. There is no doubt, however, that financiers, rating agencies and specialist financial structuring lawyers are no longer accepting insurance at its face value. Contracts are scrutin-ized as never before and increasingly insurers are being asked to sign waivers that remove or reduce their common law rights conferred under the principles of *uberrima fides*, indemnity etc. This is leading to certain insurers seeking to employ more skilled people who are familiar with financial transactions and the process of working with rating agencies, credit committees and the like to understand and, where possible, meet their demands.

Two of the key issues that affect insurance are 'certainty' and 'documentation'.

### *Certainty*

Following the film finance securitizations of Hollywood Funding one rating agency in particular, namely Standard & Poor's, recognized that relying solely upon the Financial Strength Rating (FSR) typically awarded to an insurer was not sufficient in a securitization where a multi-line as opposed to a monoline insurer is providing the insurance enhancing or supporting the issue. As a consequence they introduced the concept of the Financial Enhancement Rating (FER) which was designed to signify that a multi-line insurer supporting a particular securitization transaction had given an undertaking to 'pay first and ask questions later'. The undertaking concerned has no legal foundation and in practice would be very difficult to enforce legally, however Standard & Poor's believe that an insurer who enters into an FER agreement will not choose to ignore it because of the ramifications that such action might have on their FSR rating and reputation. The concept that in a complex financial transaction such as a securitization an insurer can, in fact, conduct a due diligence review and thus should waive the common law principle of *uberrima fides*, at least as far as the party relying upon the insurance contract is concerned, is an interesting one. At a Standard & Poor's seminar I once asked the representatives who addressed this idea if they had considered expanding the FER principle into other areas such as operational risk insurance for banks. Their view was that, in principle, this could be achieved but the amount of work involved in securing FER status for an insurer or particular insurance contract supporting a securitization process was such that it might not be practical to go beyond this area of activity for the immediate future.

### *Documentation*

Each insurance contract is a separate legal agreement between the parties that have entered into it. It is true that some standardization does exist, e.g. Form 24 – the Banker's Blanket Bond (covers a menu of crime-related risks) issued by commercial insurers to most banks that are insured by FDIC in the US. The problem is that many non-standard coverages, such as those often required by banks in support of financing transactions, are usually the subject of individual contracts. The ultimate test of an individual contract is litigation and this makes the whole process uncertain from the outset. Certain precautions can be taken in such cases, such as legal opinions on the insurance contract language etc., but this does not always reassure the financiers. If the insurance industry wishes to play a more significant role in supporting financial transactions, some form of documentary standardization needs to be achieved. The International Swaps and Derivatives Association Inc. (ISDA) Master Agreements have governed many of the standard terms and agreements that are employed every day in the capital markets. This concept of standardizing what can be standardized in agreements and thus leaving the negotiations to economic issues such as price and collateral etc. makes the capital markets more efficient and responsive. This same concept could easily be applied to the insurance industry, which seems to be determined to reinvent the wheel hundreds of times a day all over the world and mainly uses the legal system to develop its artificial intelligence. An industry body or a regulatory body, such as Lloyd's of London, could easily create a register of standard policy forms employed in the industry and develop a master database from where such standard forms could be secured in a 'registered' form, i.e. beyond tampering. At the same time this body could begin to develop industry-wide forums to agree upon standard insurance contract terminology and to provide alternative dispute resolution forums that could consider insurance terminology disputes rather than the current costly and random use of litigation.

In the world today change is much quicker than at any time in the past. Insurance, by its nature, is a responsive industry and therefore needs to keep in tune with its customers. As one door closes another door opens and while certain of the larger multinational insurance buyers are no longer as dependent upon insurance, by contrast many banks are becoming more aware of the operational risks inherent in some financial processes and they look to insurance as a mechanism to transfer such risks. The new generation of prospective insurance clients demands much more from the insurance industry. They will not accept the standard off-the-shelf products without question, rather they prefer to start by identifying the risks and then finding an insurer who understands the risks in question and can develop a risk transfer contract that will assume all or some of the identified risks with certainty. This is a problem for larger insurers since the trend that they are following is to create standard 'products' around tried and tested actuarial models, e.g. motor, homeowners etc. insurances, and then develop major marketing networks or direct internet-based marketing programs to sell these 'products'. More standardized 'product' and less individual underwriting 'solutions' seems to be the trend. By contrast many potential insurance buyers are seeking more individual solution-based insurance structures and they become increasingly frustrated when these cannot be found. Insurance is a business that has weathered many storms, both literal and metaphorical, over the past three centuries and I have no doubt that it will survive another three centuries provided that it keeps changing its role to meet the needs of each new generation of prospective clients.

# Notes

1 Certain insurers are also offering such products but the product model is more akin to banking than insurance.
2 'Cost of risk' in this context means the financial impact of expected and unexpected losses. The assessment needs to take account of the difference between pooling of losses and fiscal spreading of losses. In practice where a loss, expected and unexpected, is predicted to occur less than once in ten years, pooling of loss using a commercial insurance company is often the best solution irrespective of all other considerations.
3 'Insurable risk' in this context does not mean only risks that are insured by the commercial insurance market at the moment but, rather, it means any risk that is suitably managed via the insurance process.
4 Bank for International Settlements, Risk Management Sub-group of the Basel Committee on Banking Supervision, paper published September 1998 entitled 'Operational Risk Management'.
5 Bank for International Settlements, Basel Committee on Banking Supervision, paper published January 2001 entitled 'A New Basel Capital Accord'.
6 Bank for International Settlements, Basel Committee on Banking Supervision, paper published June 2004 entitled 'International Convergence of Capital Measurement and Capital Standards'.
7 Legal risk includes, but is not limited to, exposure to fines, penalties, or punitive damages resulting from supervisory actions, as well as private settlements.
8 Bank for International Settlements, Basel Committee on Banking Supervision, paper published April 2003 entitled 'Third Consultative Paper on the New Basel Capital Accord'.

# 6  Transferring insurable risk

*Oliver Prior*

The basic process of risk management follows the train that risk should be identified and then either eliminated or reduced and any residual risk should be transferred. Since it is very difficult in any business to completely eliminate all areas of risk it follows that there are a large number of 'residual' risks that need to be considered for transfer in one way or another.

There are many standard risk transfer contracts offered by both the capital markets and the insurance industry and in broad terms they divide as follows:

Capital markets

- Credit-related products (securitizations etc.)
- Market hedges (foreign exchange, interest rate etc.)

Insurance

- Protection of assets (fortuitous perils)
- Legal liabilities

Traditionally, capital market products tend to be in areas where considerable data are available for analysis and any assumed risk is taken on the basis of a very detailed risk analysis. By contrast the insurance industry is often willing to venture into areas where the capital markets would consider there is insufficient data available to analyse the potential for loss. Insurers are often asked to provide legal liability protection for their clients in respect of a pending piece of legislation or change in case law. Clearly there can be no data available to support the underwriting analysis of such a potential risk, yet, time after time, insurers plunge into such new areas.

Over the last two or three hundred years the insurance industry has been the main provider of risk transfer products and solutions for most corporate entities and, as such, has been relied upon to provide this service. In most corporate organizations Treasury and Insurance are seen as separate functions albeit often reporting into the same director. Treasury tends to work with capital market products, such as derivatives, and the insurance units tend to work with insurance as the main risk transfer tool. There are few people in the corporate arena who have had equal exposure to both disciplines and this tends to continue the separation of the two disciplines. Rephrasing George Bernard Shaw it may well be said: 'Capital markets and insurance are one discipline separated only by different languages.'

In the minds of many, insurance has several advantages of which the main two are that insurance premiums are usually 'tax deductible' and insurance contracts do not yet require 'marking to market', as would for example a derivative.

## The nature of insurance

Insurance has been in existence for well over 300 years and it has served industry and commerce well throughout that period.

It is important to understand that insurance contracts and capital market contracts work in very different ways. This point is often missed by financiers who seek to transfer risks using insurance. The duties of an insurance buyer can be seen to be onerous and insurance contracts are conditional. It often appears that insurance rules are unstructured and have been created and developed by myriad case laws across the centuries. This is not strictly correct since most insurance rules and guidelines can be found in the Marine Insurance Act 1906 codified in English law. It is true that the Marine Insurance Act does only apply to marine insurance contracts but the elements of any insurance contract, such as the duty of utmost good faith (*uberrima fides*), principles of indemnity etc. are all defined in the Act and the penalties for breach are also spelled out. Most insurance buyers can understand the principle of indemnity, e.g. an insurance contract is supposed to place the insured party in the same financial position after the loss as they were before it. They can also understand the point that the insured must have an insurable interest in the item damaged in order that they can be paid out under an insurance contract. Neither of these features is a requirement for a capital market contract such as a currency option or similar derivative but this is only a small matter since most insurance purchasers will have an insurable interest in the goods that they are insuring and they will be satisfied with indemnity if a loss occurs.

Utmost good faith is the issue that is most misunderstood and in many ways it is the most important feature in any insurance contract since a breach of this principle has only one remedy at law, namely voidance of the insurance contract. The Marine Insurance Act states in Section 18:

> The assured must disclose to the insurer, before the contract is concluded, every material circumstance which is known to the assured, and the assured is deemed to know every circumstance which, in the ordinary course of business, ought to be known by him. If the assured fails to make such disclosure, the insurer may avoid the contract.

The Act goes on to state: 'Every circumstance is material which would influence the judgement of a prudent insurer in fixing the premium, or determining whether he will take the risk.'

The historic logic for the duty of utmost good faith is quite clear in that it enables an insurer to enter into an insurance contract in the belief that his insured has told him everything that he needs to know about the risk he is assuming and if he has failed to do so then the insurer can avoid the contract. In the case, for example, where an insurance buyer tells his insurer that he has fitted burglar alarms to his premises where he has not, then if a burglary occurs and the insurer, in the course of investigation, finds that alarms are not fitted, then they can avoid the insurance contract. This seems fair but if in the same case the premises were destroyed by fire, i.e. the cause of loss was unrelated to the breach of utmost good faith, then should the insurer still be able to avoid the contract? The Marine Insurance Act says 'yes' since the lack of alarms *would influence the judgement of a prudent insurer in fixing the premium, or determining whether he will take the risk* and this renders that insurance contract avoidable. Of course the insurer does not have to avoid the contract but they have the right to do so and this leads to 'uncertainty'.

Legally it is possible to find ways around the duty of utmost good faith by incorporating waivers into the contract of insurance, but many insurers do not like this on commercial grounds and they consider that the amount of 'due diligence' this imposes upon them is disproportionate to the premium income related to any one risk.

Many parties seeking insurance enter into an insurance contract in the belief that they have achieved total transference of risk. In my experience, except on rare occasions, complete transfer of risk is not possible utilizing insurance even if the insurance contract contains all of the waivers recommended by the lawyers involved. For this reason two things need to be taken into account when an insurance contract is being considered as a risk transfer tool:

1   what is the reputation of the insurance counter-party, e.g. how often do they go to court to enforce their rights? (insurance would appear to be biased in favour of the insurers so this is an important issue); and
2   has any allowance been made for managing any residual risk associated with transferring risk via an insurance contract?

Insurance is an excellent tool and it has served industry and commerce well for 300 years. Today a new generation is seeking to utilize insurance to transfer risk and whether these are financiers or regulators it is important that they understand both its strengths and weaknesses when considering the role it will play in any given situation.

## Alternative risk transfer

The term 'alternative risk transfer' (ART) evolved in the late 1980s as a generic heading under which a number of separate disciplines could be grouped. The common feature of most ART disciplines is that they owe more to capital market/banking concepts that they do to insurance and, in general, they either involve reformatting a banking concept into insurance, as is the case with risk finance insurance products, or, alternatively, involve the transferring of insurance risk to the capital markets. Many of the products and experiments involving ART have been developed by individuals who were initially trained in capital market/banking skills and have transferred or been recruited into the insurance industry.

It is not my intention in this chapter to review every facet of ART in detail because some of the more experimental products have failed to gain acceptance with the ultimate buyers of risk transfer, however it is worth briefly reviewing some of these to consider why they failed and what can be learned from this.

The main components of ART that I shall focus on are:

*   risk finance;
*   securitization of insurable perils;
*   weather derivatives.

## Risk finance

In the previous chapter, 'Risk transfer in a changing world', the main difference has been considered between conventional insurance and risk finance techniques. Risk financing techniques, outside of Captive Insurance Companies, have their origins in the reinsurance industry where, in the 1960s and 1970s, a wide range of 'time and distance' contracts were developed, specifically with Lloyd's Syndicates in mind, by specialist reinsurers such as Centre Re of Bermuda. Such contracts tended to assume timing risk or provide facilities whereby insurers could establish non-specific reserves and, in practice, involved very little risk transfer. The majority of the original practitioners came from a banking background and realized quite quickly that contingent finance techniques could be considerably enhanced if they bore an insurance label.

The first significant application of a risk finance technique in the direct insurance arena was in the mid-1980s when the US insurance market virtually withdrew from providing risk transfer insurance contracts for Directors' and Officers' Liability Insurance for large commercial organizations. Since there was a need for an insurance product to be provided by a commercial insurer and self-insurance was not a viable option, Berkshire Hathaway created a risk finance contract that qualified, by the accounting rules of the day, as an insurance policy.

These programmes were generally 'one size fits all' and worked as follows:

### Directors' and Officers' Liability Insurance

| | |
|---|---|
| **Period:** | Five Years |
| **Sum insured:** | $10,000,000 |
| **Deductible:** | As agreed (but usually zero) |
| **Premium:** | $7,000,000 payable at inception |

Under the terms of the insurance contract the insurer deducted a management charge of $1,000,000 from the $7,000,000 and invested the remaining $6,000,000 in accordance with an agreed investment plan between the insured and themselves in an 'Experience Account'.

At the end of the five-year period, if the insured had reported no losses, the $6,000,000 plus interest contained in the Experience Account would be returned to them. Should a loss be payable, the amount due would first be deducted from the funds in the Experience Account (being a maximum of $6,000,000 plus accrued interest). If it exceeded this amount no additional premium would be payable but no funds would be left in the Experience Account to be returned at the end of the contract. This is shown in Table 6.1.

This method involved the minimum amount of risk transfer, i.e. 30 per cent, and relied upon the fact that significant Directors' and Officers' Liability claims take an average of seven years to mature. Should a total loss materialize, the insurer would benefit from the premium plus the interest earned prior to any payment being made that was contained in the Experience Account. If this was insufficient to cover the amount finally paid then the shortfall could be made up from the management charge levied on policies where no claims became payable. The insurer used 'time

*Table 6.1* Financial summary

| | *No claim* | *$1,000,000 claim* | *$10,000,000 claim* |
|---|---|---|---|
| Premium ($) | 7,000,000 | 7,000,000 | 7,000,000 |
| Management charge ($) | 1,000,000 | 1,000,000 | 1,000,000 |
| Investment fund ($) | 6,000,000 | 6,000,000 | 6,000,000 |
| Interest* ($) | 1,650,000 | 1,650,000 | 1,650,000 |
| Claims ($) | Nil | 1,000,000 | 10,000,000 |
| Experience account ($) | 7,650,000 | 6,650,000 | Nil |

* Interest is calculated at theoretical 5 per cent per annum.

and distance' techniques to finance the risk exposure. In the climate of the mid-1980s this type of insurance contract was very popular and over 50 contracts were bound in a period of 12 months.

From this very simple but effective insurance product sprang a whole range of risk finance insurance contracts and although it is becoming increasingly difficult to keep ahead of changes in accounting rules designed to clarify what is and what is not insurance, an increasing number of direct insurance buyers are seeking to put in place risk finance contracts to address uninsured or uninsurable areas of risk.

In order for a risk finance contract to qualify as insurance it must meet the following criteria:

1   There must be at least 10 per cent transfer of risk and the 10 per cent being transferred should have a 10 per cent chance of loss. In the US the 10 per cent transfer of risk must relate to actual risk whereas in the UK an element of timing risk can be taken into account when arriving at the 10 per cent.
2   The funding and risk transfer portions must be incapable of bifurcation. If this is not the case then the portion of the premium that is deemed to relate solely to funding will not be allowed as insurance premium for tax calculation purposes.

The formula for any risk finance insurance is:

Nett present value (NPV) + Risk transfer (RT) = 100 per cent

### Nett present value

This technique, otherwise known as 'time and distance', involves two alternative methods: pre-loss funding and post-loss financing.

In the case of pre-loss funding the insurer creates an 'Experience Account' out of which claims can be met and the object of the fund is that the premiums paid plus the investment income accrued will equal the amount payable under the funded portion of the plan (Figure 6.1). Thus, if the insurer's actuarial assessment is that a liability loss will arise after 36 months and then it will take 24 months to mature from an allegation to payment then the insurer knows that after 60 months an amount equal to the funds contribution to the settlement will need to be in the Experience Account. If the interest rate is 6 per cent compound per annum then the premium that needs to be paid to achieve the fund is 74.726 per cent of the amount required in the Experience Account to fund the expected loss. The insurer carries the 'time and distance' risk, i.e. that the expected liability loss will arise before the 36-month due date assessed by the actuarial study or that it will settle quickly.

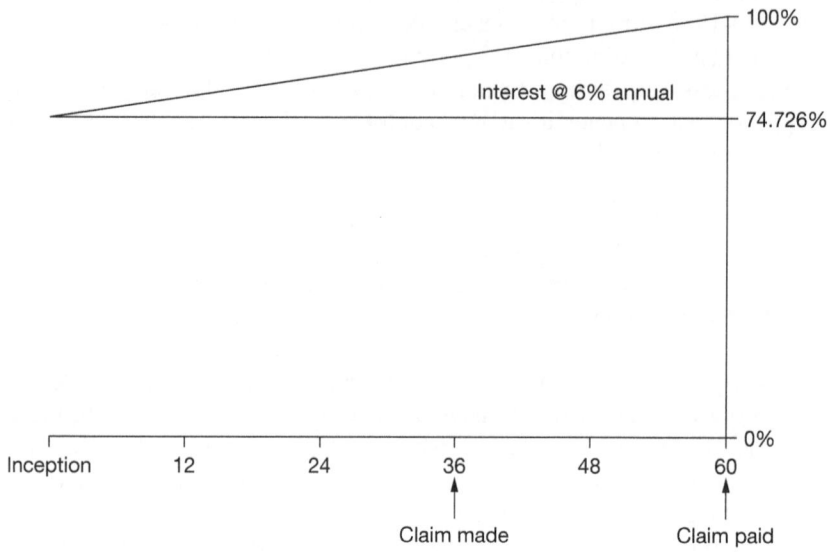

*Figure 6.1* Nett present value

Post-loss funding involves the insurer making a payment and the insured then repaying this over a period of years. Some post-loss plans involve interest payments to achieve 'nett present value' and compensate the insurer for loss of use of the fund and others do not include interest in the repayment formula. The incorporation of interest in the repayment structure could cause the repayment premiums to be questioned by the Inland Revenue.

### Risk transfer

Risk transfer is the insurance part of the plan where a premium is paid to purchase a level of risk transfer from insurers or the capital markets depending upon the type of risk. The premium rates applied will vary according to the risk exposure.

## Securitization of insurable perils

Catastrophe-linked securitizations have come a long way since the first issues in 1996 that sought to transfer natural catastrophe risk to capital market investors. In the early days many different techniques were applied to test investors' appetites. Initially some issues sought to convert the unfamiliar concept of natural catastrophe to the more familiar concept of credit risk by seeking to construct contingent lines of credit triggered by 'events'. In a short period of time however it was proven that natural catastrophe risks could be transferred to capital market investors provided the price is right. Again, early issues sought to combine loss of principal with guaranteed principal but loss of 'coupon' following an event. Despite the rating differences for guaranteed principal to loss of principal, the loss of principal issues quickly became acceptable provided the pricing reflected the perceived risk.

In his excellent paper entitled 'The Perfume of the Premium' Morton Lane[1] quotes the old adage of the capital markets, 'the perfume of the premium overcomes the stench of the risk' – in insurance people are more familiar with the term 'greed over discretion'.

There have been many excellent books and articles written about the subject of catastrophe-linked securities that analyse each issue. My aim is only to review overall developments and, in particular, prices. This review will consider:

- structures;
- loss assessment methods;
- capacity;
- prices.

### *Structures*

The basic structure of a catastrophe-linked securitization has now been established and it is shown in Figure 6.2.

While there may be variations it is likely that most structures will follow the pattern illustrated. One thing is becoming apparent, namely that if more investors are to be attracted to supporting this class of security it will need to provide a good risk spread. This means that more new issues will be required to create a true risk spread. Re-creating a structure each time will not induce more investors, rather, it will have the opposite effect. Hence the desire to create a 'formula' for the structure that can be instantly recognized by any potential new investor.

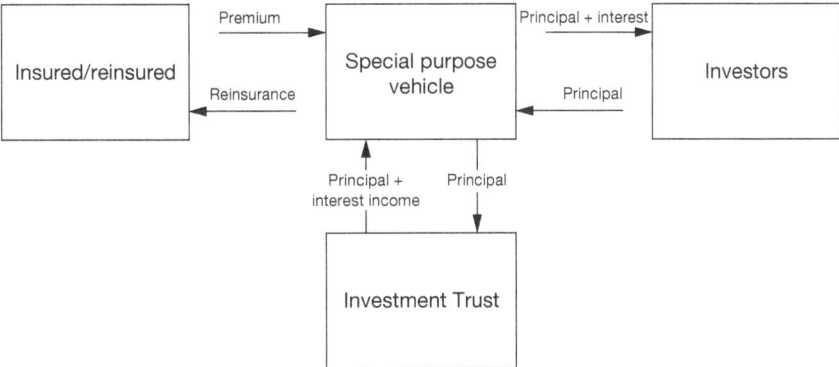

*Figure 6.2* Risk securitization structure

### Loss assessment methods

There are three generic methods for assessing loss under natural catastrophe linked securities:

* *indemnification* – where the loss matches the loss sustained by the 'issuer' of the security;
* *index linked* – where the amount payable following an 'event' is calculated by reference to an index that tracks the possible loss causes that might be suffered by the security 'issuer'; and
* *parametric* – where the amount payable is calculated by linking a specific payable amount to a measurable event such as an earthquake measured on the Richter scale or a hurricane measured on the Saffir-Simpson Scale.

At this point it is important to consider the issue of 'basis risk' both in general and specifically as it relates to catastrophe-linked securities.

### Basis risk

Basis risk is the difference between the amount of loss that is actually suffered by an organization or insurance company and the amount that is collected by way of reimbursement under a 'capital market' contract.

Basis risk illustrates one of the main differences between capital market and insurance 'products'. Insurance policies that provide for damage are designed to be specific and it is a requirement of an insurance contract that it should provide indemnification, i.e. the purchaser should be placed in the same position after a loss as they were prior to it and it is not permitted to make a gain on a contract of insurance.

*Indemnification*, *insurable interest* and *uberrima fides* are all principles enshrined in the insurance process. In order to provide a comparable system the capital markets have had to address these issues.

Indemnification provides for reimbursement of the actual loss sustained whereas index-linked and parametric structures are less likely to provide indemnification.

One of the major differences between the 'products' offered by insurance and the capital markets is that insurance was developed to address a client's specific needs whereas a capital market instrument is usually designed with one eye on an ability to 'trade' it on a secondary market. An ability to 'trade' a 'product' on a secondary market means that it cannot be too specific. By way of illustration the issuer cannot transfer an indemnification catastrophe-linked security, however it is possible to transfer a parametric-style bond. It would theoretically be possible for an issuer of a parametric trigger catastrophe-linked security to transfer it across to another reinsurer seeking similar triggers. Or, as has been the case, for piggyback securities to be issued that reflect the triggers of an existing security.

One of the main issues relates to modelling. Since the capital markets will require 'due diligence', e.g. modelling, and the ultimate securities will need to be rated by one of the rating agencies, a variety of information will be required for this purpose. The information divides into two broad categories:

- catastrophe event specific;
- portfolio specific.

Catastrophe event-specific data relate to the event being linked to the issue and the models will seek to show the general impact of varying degrees of that event on the region being considered. Such data will be more 'publicly' available than portfolio-specific data. Portfolio-specific data show how the event being modelled will impact upon a specific insurer's portfolio in the region being considered. The degree of information required for each issue type is as follows:

Indemnification   =   Catastrophe event specific + Portfolio specific

Index linked   =   Catastrophe event specific + an Index

Parametric   =   Catastrophe event specific

Despite the fact that the modelling associated with indemnification is more of a hassle for the issuer, the majority of issues in 1999 and during the early days were on this basis. Today, more issues are being structured around indices or parametric loss triggers to speed up the issue process and control the costs.

Firms that can master basis risk in one structure will undoubtedly be able to offer the most flexible 'products'. Lehman Brothers underwrote US$150,000,000 of California Earthquake insurance in their Bermuda-based insurance vehicle Lehman Re and then 'transformed' this purchasing and index-based 'catastrophe linked security' that created a partial hedge leaving the basis risk to be retained in Lehman Re. Lehman maintains that according to their models the basis risk retained by them in this structure is minimal and the premium that they are retaining is more than sufficient to reward them for this.

There is no doubt that the buyers of catastrophe insurance don't like basis risk. Equally the capital markets prefer to accept risk on a parametric or indexed structure and accordingly will offer benefits in the 'spreads' for doing so. Lehman has exploited this feature to arbitrage the two positions to their advantage. This has shown that in catastrophe insurance the ability to 'transform' will be critical to future success.

### Capacity (liquidity)

The lead management of insurance-linked securities has tended to be limited to a few major Wall Street firms that have the capability to 'place' the securities. There is a major appetite for such instruments in the insurance community, however, the managers have tended to restrict the sale of such issues to insurance companies directly in order to encourage other investors. There are a few dedicated funds targeted solely at catastrophe-linked investments. Such funds illustrate that several investors have an interest in this new class of security or investment but do not find it sufficiently mature to warrant investment in employing 'in house' expertise that understands catastrophe insurance and capital market practices. Many issues are over-subscribed which indicates a surplus of capacity but the number of issues is too low for the experts to assess market capacity on a daily basis. The insurance market benefits from an enormous deal volume that allows 'managers', such as insurance brokers, to benchmark risk capacity on a daily if not hourly basis. It will be many years, if ever, before the capital markets can provide such accurate capacity forecasts.

The secondary market is holding up well and trading is quite robust for the following reasons:

1   insurance companies that were restricted first time around see the secondary market as an entry point;
2   prices of the earlier issues have been reduced on a benchmark basis by subsequent issues and this indicates better value in the older long-term issues.


### Prices

It is very difficult to benchmark prices in a market where issues are rare and are not always comparable. One exception to this is the Residential Re/United Services Automobile Association (USAA) issue. USAA attempted the first issue in 1996 and failed to complete the placement but since then they have completed seven catastrophe-linked securitizations and the changes in these issues over the seven-year period reflect many of the changing trends in the catastrophe-linked securities markets.

USAA is a mutual insurance company that underwrites personal lines insurances for US military officers and their families. Its premium income is around $5 billion derived from approximately 2.5 million policyholders located across the US. USAA's book of business reflects the demographic trends of the retired population of the US; namely, it is becoming concentrated in catastrophe risk areas such as Florida and Texas. USAA has less ability to control its catastrophe exposure because of its target market.

In 1996 USAA tried to bring a catastrophe-linked issue to the market but was unsuccessful. The issue was for $500 million and came at a time when capacity for such issues was around $100 million. In 1997 USAA were back again with a similar sized offering and were successful, and they have returned again each year through to 2003 so far.

The history of the offerings is as follows:

### USAA/Residential Re

| | |
|---|---|
| Ceding Company | USAA |
| Security Issuer | Residential Re |
| Period | One year but increased to three years from 2001 issue (cumulating) |
| Loss Assessment | Indemnification |
| Geographic scope | Various US Eastern Coastal States and District of Columbia but increasing to include, for example, Hawaii and other exposures as the issues changed over the years. |
| Risk analyst | Applied Insurance Research |
| 1996 issue | Withdrawn |
| 1997 issue | Placed US$477,000,000 Average 'risk price' 576 bsp (over LIBOR)[2] |
| 1998 issue | Placed US$450,000,000 Average 'risk price' 400 bsp (over LIBOR) |
| 1999 issue | Placed US$200,000,000 Average 'risk price' 366 bsp (over LIBOR) |
| 2000 issue | Placed US$200,000,000 Average 'risk price' 410 bsp (over LIBOR) |
| 2001 issue | Placed US$150,000,000 Average 'risk price' 499 bsp (over LIBOR) |
| 2002 issue | Placed US$125,000,000 Average 'risk price' 490 bsp (over LIBOR) |
| 2003 issue | Placed US$160,000,000 Average 'risk price' 495 bsp (over LIBOR) |

The expected annual loss figures[3] provided by modelling were:

| | |
|---|---|
| 1997 | 0.63% |
| 1998 | 0.58% |
| 1999 | 0.45% |
| 2000 | 0.54% |
| 2001 | 0.68% |
| 2002 | 0.67% |
| 2003 | 0.48% |

If the 'risk price' is adjusted to reflect a level risk then the risk price becomes:

| | |
|---|---|
| 1997 | 0.914 bsp per 0.01% of risk |
| 1998 | 0.690 bsp per 0.01% of risk |
| 1999 | 0.813 bsp per 0.01% of risk |
| 2000 | 0.759 bsp per 0.01% of risk |
| 2001 | 0.734 bsp per 0.01% of risk |
| 2002 | 0.731 bsp per 0.01% of risk |
| 2003 | 1.031 bsp per 0.01% of risk |

Equally the market pricing can be viewed across three different issues as shown in Table 6.2.

While these issues contain a number of differing features it is possible to conduct a 'rule of thumb' comparison across all three.

It is very difficult to read anything into the above pricing data other than the fact that they form a good basis for understanding catastrophe risk pricing in the capital markets. Morton Lane of Lane Financial LLC[4] based in the US has published a number of excellent papers concerning insurance securitization and price trends and co-relations. I would recommend anyone who is interested in reading more about this subject to visit the website www.LaneFinancialLLC.com where they will find a number of excellent technical papers that can be downloaded free of charge.

A major concern among market makers is the narrow spread of the issues currently on offer. Catastrophe-linked securities are still expected to 'trade' at a premium over their equivalent rated corporate bonds. One explanation for this is that with corporate bonds of equivalent rating an investor can reasonably expect some recovery (averaging 40 per cent) and such recovery is not possible in the current catastrophe-linked securities issues. Will this trend continue or will the catastrophe-linked securities fall to a pricing level that is comparable to corporate paper. I suspect that answer is 'chicken or egg' and, as so often occurs in such cases, an entirely unrelated event may cause a breakthrough.

Various attempts have been made to apply the securitization structure developed for the transfer of natural catastrophe risks to other insurable

*Table 6.2* Market pricing

| Year | Trinity 3/98 | Residential 10/99 | Juno 6/99 |
|---|---|---|---|
| Risk factor (%) | 0.77 | 0.44 | 0.45 |
| Risk premium (bsp) | 417 | 366 | 420 |
| Premium/risk factor | 5.4 | 8.3 | 9.3 |

risks, such as aviation accidents etc. but these have mainly been restricted to small private issues of US$50,000,000 or lesser amounts. With one notable exception to date all major 'public' insurable risk securitizations have been in respect of natural catastrophe risks. The notable exception is the Golden Goal Finance Limited SPV risk transaction arranged in 2003 that will provide FIFA with €400,000,000 protection in the event that the 2006 World Cup event, to be held in Germany, has to be cancelled due to war, boycott etc. This latter transaction is the first 'public' securitization that involves the process of transferring 'human risk' as opposed to natural catastrophe risk to capital market investors.

Up until the Golden Goal Finance transaction it would have been safe to say that there was no appetite in the capital markets for 'human risk' except where very small private offerings are involved. Golden Goal has broken new ground and one of the main reasons for the success of securitization was the detailed risk analysis carried out by Risk Management Solutions (RMS). Golden Goal Finance has shown that with the right data available 'human risk' can be transferred to capital market investors.

There is considerable need for risk transfer where the risk involves human acts, e.g. fraud, error etc. The insurance market for such risk is restrictive in the cover that it provides and the limits of indemnity available are often inadequate for the size of risk faced by prospective buyers. Expanding insurance-linked securitizations into the area of human risk-related perils could increase the demand for such issues considerably.

## Weather derivatives/insurance

Weather derivatives and related insurance products were pioneered in the US. Over the last few years there has been increasing interest in this cover in both North America and the EU. The weather product that has been developed is capable of being provided as a derivative or an insurance contract in virtually identical forms and this makes it unique.

Adverse weather can have a severe impact on the trading results of companies in a number of sectors, including utilities and retail. In recent years more than one UK-based retailer announced disappointing results due to 'unseasonable' weather during the summer months.

The issue relating to weather insurance was a common one for the insurance industry, namely, an insurance contract normally requires two key components:

- insurable peril;
- a method to quantify loss.

The insurance industry had had no problem insuring weather-related perils and everyone is familiar with the insurance purchased by organizers of fêtes and similar outdoor events to protect them specifically against the loss that they face (in such cases the loss is normally the cost of arranging the event etc.) as a result of the event being cancelled due to adverse weather. A large retailer or supplier of seasonal products such as ice cream, however, could not purchase an insurance product that specifically paid them should they face adverse weather conditions across a season or specific period of time, e.g. bank holiday.

The weather derivative and the subsequent related insurance contracts sought to address the issue of linking adverse weather conditions to an index in order to provide a means of measuring whether an 'insured event' had occurred and, if it had, fixing the payment. This use of mixing a capital market technique, e.g. index linked structures, with traditional insurance perils created an exciting new risk tool and showed the way for future similar constructions.

### *Creating the index*

In order to quantify the risk exposure it is first necessary to create an 'index'. In the illustrated example, this is achieved by determining a mean average 'benchmark' annual temperature across North America, this being 65 degrees Fahrenheit.

Computer models are then established that determine whether each particular day is a 'Heating Degree Day' (HDD) or a 'Cooling Degree Day' (CDD). A 'Heating Degree Day' (HDD) is determined to be one where the average temperature will be below 65 degrees Fahrenheit, i.e. where it is deemed that people will wish to heat the ambient atmosphere. A 'Cooling Degree Day' (CDD) is determined to be one where the average temperature will be above 65 degrees Fahrenheit, i.e. where it is deemed that people will wish to cool the ambient atmosphere.

The actual HDDs and CDDs are stated in numeric form by taking the number of expected degrees on each day above or below the benchmark. Thus, on 28 August if the average expected temperature calculated by the models across the modelling period is, say, 85 degrees Fahrenheit, 28 August will be a CDD (because cooling of the ambient atmosphere will be necessary) and it will be accorded a value of 20 (being 85 − 65). An average climatic model is then built using HDDs and CDDs (see Figure 6.3).

With this model in place it is possible to design a derivative or insurance contract, depending upon the client's needs or wishes, which would respond should there be an actual variation in the average HDDs or CDDs across a given period when compared to the 'index'.

A value is then placed on each unit of HDD or CDD and variations from the norm to determine the amount paid under the insurance or derivative contract.

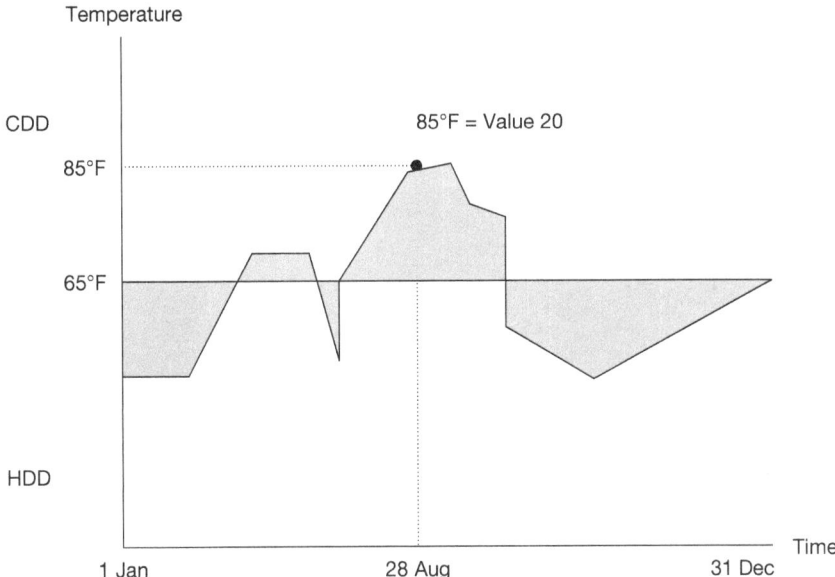

*Figure 6.3* Example HDD/CDD weather model

### Example

XYZ is a utility that has determined the estimated amount of natural gas that it requires to purchase in an ordinary year and it has entered into 'fixed price' purchase agreements at favourable terms to ensure the best cost base (see Figure 6.4). If the actual number of HDDs is outside the norm then it is likely that costs will be incurred in buying in additional natural gas supplies at 'spot' prices. XYZ determines that for every HDD outside the budgeted norm there is a potential additional cost of US$5,000, being the difference between the fixed cost purchase price and 'spot' for natural gas.

Thus, weather derivative/insurance is arranged on an annual basis as follows:

| | |
|---|---|
| Sum insured: | US$5,000 each HDD outside the index average across the 12-month period |
| Deductible: | 200 HDDs |

If during the course of the year it is determined that the expected average HDD value is 5,000 and during the period the actual is 6,000 then the payment made by the insurer will be 1,000 HDDs less 200 HDDs deductible, i.e. 800 HDDs, and since each HDD has a value in the derivative/insurance of US$5,000 the market will pay US$4,000,000 (being US$5,000 × 800).

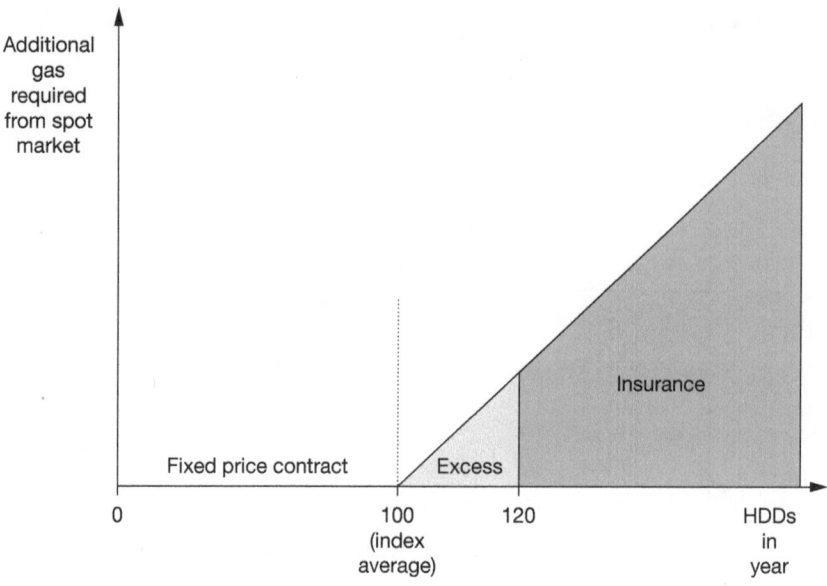

*Figure 6.4* Illustration of the application of a weather derivative contract

### Application

Weather derivative/insurance contracts are very dependent upon statistical information and in the US the National Weather Bureau's statistics are used to determine the actual value of CDDs and HDDs. In other countries, such as the UK, sophisticated data are available but may not be as 'local' as a buyer may require. For example, the weather station at Heathrow Airport provides an excellent statistical basis for modelling but this ties the purchaser into that weather station locality – this may not have application if, say, the buyer is seeing their main exposure as being in Reading.

Weather derivative/insurance programmes are also being used in the US to construct a new form of 'fixed cost' electricity supply contract to major stable consumers such as commercial offices etc. Fixed price contracts are offered at a 'premium' and the premiums are then transferred to an insurance vehicle (captive, SPV etc.) where they are used to purchase reinsurance or derivative products to hedge any potential loss consequent upon the fixed price contract compared to the floating rate version.

The problem with many conventional financial loss insurance contracts is that in order to be indemnified for financial loss a physical loss must occur and causal link must be shown between the physical damage and the financial loss. Such financial loss insurance is even named consequential loss insurance in order to show that it is related to another event such as physical damage. A good example would be an hotelier whose business is impacted by 'foot and mouth' disease. The cattle affected are not owned by the hotelier so he cannot insure the animals and thus any consequential loss stemming from any disease affecting them cannot generally be insured against. It is theoretically possible to develop an index-based product that could provide protection for anyone from potential financial loss that they may face as a consequence of a 'foot and mouth' outbreak. The slaughter of animals as a consequence of a 'foot and mouth' outbreak is very carefully controlled by the government and, using official data, an index of animals slaughtered in each year could be developed. In a manner similar to the weather derivatives/insurance, financial loss could be linked to the index. In this way it would be possible for an hotelier to purchase 'points' on the index and receive payment that would increase in accordance with the number of animals slaughtered. This would not, of course, provide true indemnity but it would provide a quick means of compensation for hoteliers faced with such an event.

The weather derivative/insurance product has served two purposes, namely it has provided valuable protection to certain organizations that hitherto could not find such protection and it has shown how a capital market technique such as indexation can be harnessed by insurers to build a new and more responsive set of financial loss products.

## What can be learned from recent 'experiments'

The last decade has seen a revival in experimentation in new ways to transfer risk. In the 1960s and 1970s many bank groups acquired interests in insurance broking firms because they saw potential for synergies in the large corporate account arena. Most of these dreams were never realized and by the 1980s it was no longer fashionable to talk about Bancassurance in the context of corporate clients. Instead, the very banks that had made the investments in the early decades into insurance brokers that specialized in the corporate sector, began to offer personal and SME-related insurance services on a direct basis. They found that both as principal and intermediary simple insurance products such as motor and household could be sold directly to their client base with relative ease. In the 1980s, when the UK motor and household insurance market was dominated by names such as Royal Assurance, Guardian, Norwich Union etc., had anyone suggested that in less than 20 years the second largest insurer in this section would be a bank-owned 'start-up' they would have been considered mad or at best deluded – yet this is just what happened. By contrast the major corporate clients resisted the idea of 'one stop shopping', rather they sought to engage an ever wider range of specialist advisers proving that expertise and not bulk buying ruled in this market place.

In the corporate sector the real benefits of banking and insurance working together began to be realized in the late 1980s and 1990s. The initial changes came about when individuals from the investment banking world were hired by specialist reinsurers to develop and promote risk finance reinsurance contracts. Specialist *time and distance* reinsurance contracts were initially targeted at the Lloyd's market where loss reserving requirements were more stringent. The combining of contingent lines of credit with insurance practice led to risk financing becoming part of the insurance 'tool kit' and ultimately to a range of 'finite' insurance products (so named because they offer, by the application of aggregate limits of indemnity, a finite level of protection during a policy period) and blended finite insurance products (so named because they blend time and distance techniques with risk transfer in order to make sure that they qualify as insurance contracts).

While innovation in risk financing was happening, by contrast, the conventional insurance industry was going through a period of introversion. Lloyd's was going through a metamorphosis and many of the insurers that had previously been risk pioneers did an 'about face' and began to consolidate their offerings to tried and tested conventional products. Lloyd's, the home of insurance innovation, was being tamed. Business plans, risk-based capital, corporate capital, professional reinsurers etc., all had the effect of controlling the more innovative underwriters. The larger insurance companies embraced alternative risk transfer and, in particular, finite and blended insurance contracts, since it enabled them to behave as though they were being incredibly creative while, at the same time, assuming very little in the way of transferred risk. Throughout the 1990s insurance buyers were being encouraged by corporate governance to go forth and find ever-newer risks inherent in their business, and, having identified them, to manage them. Since part of the risk management process is to transfer any residual risk they looked to insurance. The assumption of a 'one off' risk identified by a prospective insurance buyer was no longer something that insurers were keen on so they were referred to the ART department where risk finance solutions were often dispensed by way of a substitute for real risk transfer insurance. An interesting question to ponder is what if risk finance had not been invented – would insurers have sent the prospective buyers away empty handed or would they have made more of an effort to develop true risk transfer solutions to address these emerging risks?

In addition to a lack of willingness to consider the one off risk on a true risk transfer basis the insurance industry began to develop an aversion to 'systemic risk'. In some areas such as political risk and natural catastrophe risk, insurers have always been aware of the potential for a systemic loss and, as such, they monitor aggregate exposures and purchase suitable reinsurance. Other areas such as computer virus and terrorism were not seen as having the potential for a systemic loss until a number of unrelated events such as the bombing of the City of London in the early 1990s, Y2K computer-related issues etc. focused attention on the enormity of systemic risk. How can you possibly underwrite computer-related risks in a world where communications has become one giant computer network and viruses can pass from one machine to another with ease? As far as the insurance industry was concerned the answer was 'you cannot' and in the run up to Y2K the majority of the insurance industry said no to any client who sought protection. Those who did offer anything did so on a risk finance basis or offered small nett line contracts at very high rate on indemnity. At this point in time any perceptive insurance buyer who had not realized it before began to recognize that the insurance industry would no longer be offering a full risk transfer service for any fortuitous event, rather, like the capital markets, it would offer a series of risk transfer products that would address the common but non-systemic severity losses.

Innovation is not dead in the insurance industry by any means. Environmental risk, intellectual property, reputational risk, unauthorized trading, computer crime etc. are among the many emerging risks where excellent risk transfer insurance contracts have been made available in recent years. 'Product' innovation is alive and well and provided that a number of potential insurance buyers face a common risk exposure it is possible for them to work either individually or together with one of the more innovative insurers to develop a true risk transfer product. The work involved in bringing a new product to market is considerable and, in addition to internal expenses, not inconsiderable external cost is likely to be incurred for legal and related expert advice. Since this cost must be recouped from the premium derived from future sales of the product, a mass sales culture as far as innovation is concerned is inevitable.

By contrast the insurance industry is no longer geared up to providing one off solutions to the large number of potential insurance buyers that are being driven by corporate governance to identify non-standard risks inherent in their business. The fixed costs associated with insurance innovation in the modern insurance world, such as due diligence reviews, legal costs for policy wording reviews and underwriting time, all conspire to make the cost of an innovative risk transfer solution too expensive for most insurance buyers.

A true 'chicken and egg' stalemate seems to have been created as far as innovative one off risk transfer solutions are concerned in the insurance industry. The potential buyers of insurance contracts are familiar with the 'no cure no fee' practice associated with insurance buying. Insurers, concerned about devoting many hours of time and expense to developing a one off risk transfer solution that may ultimately be rejected by the prospective purchaser, have tended to shun such projects in favour of product innovation. The answer would seem to be a change in the way both sides address such risks. Prospective insurance buyers need to understand that for one off solutions insurers are likely to require commitment/ abort fees to cover the costs associated with developing a one-off solution where no purchase of the end product is guaranteed. Insurers need to make sure that they can still address one off risk transfer solutions in order to meet the growing demand in this area.

## Notes

1   Paper entitled 'The Perfume of the Premium ... or Pricing Insurance Derivatives' written by Morton Lane and published by Lane Financial LLC.
2   The issue was split into two parts for 1997; part being AAA rated capital protected and part being BB with the possibility of capital impairment. It is the latter part that attracted this risk price.
3   The probability of an event causing any loss to the security protected layer.
4   Lane Financial LLC, Willmette, Illinois.

# 7 Recent risk financing innovations

## Motives, principles and practices

*Peter C. Young*

## Introduction

Innovations in risk financing have arisen from a broad range of factors that are changing the overall face of risk management. Any discussion of risk financing innovation, therefore, must be placed in this broader context.

Historically, there have been two dominant views of risk management. The first view, often called 'traditional risk management', is defined as insurance buying and the management of conventionally insurable risks (through safety programmes and other measures to control such risks). The alternative to this view was first established in financial institutions where risk management was defined as the coordinated management of financial risks (interest rates, currency, price and credit, for example), a view known as 'financial risk management'. Both views have broadened over time and overlapped to a certain degree with the result that some links have been established in recent years.

Independently of these histories, other forces have emerged in recent years leading to a third, and broader, interpretation of risk management. There are numerous reasons for these external forces, but suffice it to say here that legal, regulatory, internal control, managerial, and even academic factors have influenced the development of a view of risk management that is comprehensive, systematic, organization-wide and mission driven. This view has come to be known as enterprise risk management (ERM) – a subject that is covered at length in Chapter 8 by Gerry Dickinson.

It is against this backdrop that innovations in risk financing should be considered. Risk financing is illustrative of many innovations that arise when traditional and accepted practices come under pressure from forces for change. For example, some innovations in risk financing have been motivated by the desire to find new tools that can span the boundaries of historical risk management. Other innovations show a linkage between existing traditional and financial risk management tools. And, a few innovations represent forays into *terra incognita*, that is to say, areas of risk management where solutions could not be found through simple modifications of historical financing tools.

The preceding comments suggest that demand for risk-financing innovations (from buyers) is the principal motive in play. However, innovation also has arisen from competitive pressures within the supply side of risk financing markets. Many banks, reinsurance companies and investment banks developed innovations to better position themselves against competitors and to find new markets for traditional products.

There have been several larger waves of innovation over the past 30 years, with a number of smaller ripples, and one of the results has been that the terminology of risk financing innovation is somewhat imprecise. Although the chapter will attempt to briefly address this issue, the strategy will be to examine the most recent type of risk financing innovation – capital market-based products. By focusing on a specific type of innovation, a more general range of risk financing and innovation issues can be addressed.

In order to explore the subject of innovation in risk financing, this chapter asks – and seeks to answer – the following questions:

1   What is risk financing, and how has the definition changed in recent years?
2   What are the specific influences on current innovation in capital market-based risk financing?
3   Can we classify and explain the most recent approaches to innovative risk financing?
4   What are the issues, challenges and problems that attach to innovations in capital market-based risk financing?

## What is risk financing, and how has the definition changed in recent years?

Early in the development of traditional risk management the definition of risk financing was tied closely to insurance. Indeed, early risk management books did not have a descriptive term for measures that did not involve insurance, except to say that such risks were 'retained'. But even then, retention principally was described as those parts of an insured risk that were paid directly by the policyholder (deductibles, co-insurance, co-payments). Such was the focus on the management of insurable risks that the financing of non-insured risks was not much mentioned.[1]

In the 1960s and 1970s several changes began to occur that expanded this traditional view of risk financing. Increasing professionalism among risk managers and the need to respond to successive problems in insurance markets led to a period of innovation and experimentation in strategies intended to serve as alternatives to insurance.[2] Self-insurance programmes and captive insurance arrangements are examples of resulting innovations. However, while the *means* of risk financing broadened, the implicit definition of risk financing remained: *risk financing was any measure taken to finance the cost of losses or potential losses.*

At roughly the same time that traditional risk financing began changing, separate developments were occurring in financial markets. Of particular importance were the emergence of formal secondary markets and the growth of financial risk management tools – especially forwards, futures and options contracts.[3] The framework for thinking about risk financing in this context was, plainly, different. *The overall purpose of risk financing was to control or neutralize (and, sometimes, to speculate on) certain financial risks.*

Therefore, since the 1970s the definition of risk financing rested uneasily on these two different frames of reference. For traditionalists, risk financing was defined as any measure that financed the cost of losses or potential losses, whereas for financial risk managers risk financing was those measures taken to hedge or otherwise control exposures to financial risk.

By the late 1980s, an important change began to occur on the traditional side of the field. That change was a recognition that risk financing needed to include all measures intended to manage the cost of *risk* – not just the cost of losses. Risk imposes many costs on organizations: some tangible, such as a risk manager's salary or a fire-related loss, and some intangible, such as uncertainty or the misallocation of resources due to worries about risk.[4]

Change also occurred on the financial risk management side with growing interest in operational risks. Sensational cases like Barings Bank underscored the fact that while financial risks – for example, currency exchange – might be controlled with financial instruments, ultimately the 'handling' of those instruments was down to human beings and organizational processes, both of which brought non-financial risks into the equation.[5] In that light, financing of measures to control operational risks (both tangible and intangible) seemed to be part of the bigger challenge of managing financial risks. As a result, a general consensus began to emerge among traditional and financial risk managers that risk financing could/should include all financing arrangements that arise from the presence of risk.

Importantly for this chapter, the recent emergence of ERM even further expands the preceding definition.[6] As is true of the traditional view, 'making post loss resources available for reinvestment' remains one important purpose of risk financing. And, within the financial risk management framework, the desire to create organization-spanning – and consistent – financial strategies is present as well. However, ERM frames the risk financing question at a more fundamental level: what approach to risk financing will support optimal capital structure and sound potential re-investment decisions, and assure that risk costs do not negatively affect firm value? In this context, modern risk financing decisions are not *just* driven by the desire to produce resources after a loss or the need to integrate strategies and tools, but, rather, by overall capital structure, investment and re-investment, and risk reduction (or even risk taking) objectives.[7] Put this way, the ERM definition of risk financing could be, *all financial tools that specifically address risk with the objective of maximizing firm value.*

Before leaving this subject, a few additional comments regarding terminology are useful. In the early days of insurable risk innovation, the term *alternative risk financing* (or, more narrowly, *alternative risk transfer*) was applied to all non-insurance risk financing arrangements.[8] Consequently, prior to the 1990s a very great number of alternative risk financing/ transfer tools grew and developed (self-insurance programmes, captive insurance companies, group captives, finite reinsurance programmes, to name a few examples). From the traditional risk managers' perspective, this led to a problem of terminology when financial risk management tools began to enter their field of vision. Are financial risk management tools 'alternatives' as well, and if so, alternative to what? To be fair, the problem equally existed for financial risk managers encountering traditional risk financing tools and their derivations.

It is not the purpose of this chapter to fully resolve this problem of terminology. Suffice it here to say that a range of tools and products initially were conceived as direct responses to insurance availability and affordability problems, and to a desire on the part of risk managers to better manage insurable risk costs. For this chapter's purposes, these tools (captive insurance companies, self-insurance pools, risk retention groups) could be labelled *traditional innovations.* On the financial risk management side, the initial innovations (forwards, futures and options) were developed for specific financial risk purposes, and thus could be labelled *financial risk innovations.* The most current range of innovations combines aspects of both traditional and financial innovation, but applies them in an ERM context, which suggests the label *enterprise risk financing innovations.* These current innovations tend to utilize capital markets, but are not limited to capital market usage, so we conversationally refer to them as capital market-*based* risk financing innovations. It is on this form of innovation that this chapter has set its sights.

## What are the specific influences on current innovation in capital market-based risk financing?

*Traditional innovations* were driven by crisis. Successive insurance sector problems between 1950 and today pressurized insurance buyers to seek alternatives to insurance coverages. *Financial risk innovations* developed more through a general evolution of financial and commodity markets than by crisis. Experimentation with formalized tools for controlling financial risks led to the formation of secondary markets within exchanges (Chicago Mercantile Exchange, Chicago Board of Trade, for example). *Enterprise risk financing innovations* differ, in that they seem to be due to a convergence of several factors.

### Demand-side influences on recent innovation

#### The role of intermediaries

The changing dynamics of financial intermediary markets led to an environment in which investment and insurance brokers sought to expand and diversify their services, differentiate themselves from competitors, add value to their existing services and products, and to enhance their professional reputations.[9]

In the 1990s one particular way this situation influenced risk financing was the experimentation that occurred within insurance and investment brokers in designing new tools that connected traditional insurance products with investment and financial risk products. As intermediaries are, typically, agents of their customers (and not financial markets), their motives linked with developments within the risk management community.

#### External regulatory and legal pressures

In general, tax law and regulation of business cannot be said to particularly endorse innovation in risk financing. For example, tax treatment of uninsured losses is less favourable than is treatment of insured losses. However, in recent years a number of initiatives have occurred in developed economies that have emphasized the importance of coordinated and systematic approaches to managing risks. Sarbanes-Oxley hints at this, and the Committee of Sponsoring Organizations of the Treadway Commission (in its Enterprise Risk Management Framework) has now established the basis for a future standard. In the UK, Turnbull, the Combined Code, and a range of related statements have established the principle of coordinated risk management – and, indeed, similar standards have emerged in Australia, Canada, Germany, Denmark, Japan and elsewhere.[10]

The full effect of these new external expectations is yet to be realized, but an important feature of these developments is the role that external auditors will play in verifying that such risk management measures are undertaken. Since an implicit aspect of the audit process is to establish whether firms are consistent, well managed and responsive to the needs of owners and key stakeholders, there will be most likely the emergence of a risk financing frame of reference that will more directly connect risk financing measures to the value-maximizing purposes of the firm.

The timing of these recent changes means that they did not have a direct effect on innovation in the 1990s, except to the important extent that the emerging consensus represents the culmination of a process begun in the 1990s and centred on a wide-ranging discussion about the role of risk management in organizations. In that sense, the current measures ratify ideas and motives that supported the most recent wave of innovations.

*Enterprise risk management*

Related to regulatory and legal trends, the intellectual movement known as ERM must be mentioned as an influence on innovation in the 1990s.[11] As is true with regulatory and legal factors, the influence of ERM is indirect at this point in history. There is no available evidence to show that firms are innovating in risk financing as a result of their move toward ERM practices. However, the debate over the ERM concept, which began to occur in the early 1990s, provided a context in which a wider array of questions was raised about risk financing strategies. Centrally, the ERM framework posed the question, 'since integrated risk management is desirable from a broad business perspective, should not risk financing follow an integrated and organization-wide form and practice?'

*Internal managerial pressures*

Risk managers within organizations have been seeking new ways to innovate, particularly in light of pressures to (1) adopt a wider field of vision with respect to organizational risks and (2) respond to highly volatile and unpredictable insurance markets. There is a push–pull effect here, with brokers encouraging risk managers to innovate by offering solutions, but 'advancing professionalism' should not be dismissed as an influence. The central dynamic of risk management since the mid-1980s is toward broader scope, higher visibility and a desire to increase the managerial competence of practitioners. Innovation (of all kinds) has been a consistent theme in professional publications during this time.[12]

## Supply-side influences on innovation

*Competition*

Competition among and between insurance brokers, reinsurance companies, investment banks and other financial services organizations has plainly heightened motives to innovate or differentiate. Although levels of competition abated in 2000 and thereafter (due to 9/11, the downturn in investment markets and a range of political and economic concerns), the 1990s were witness to a very high level of activity in the areas of new product development, R&D and cross-sector collaboration (e.g. investment firms working with reinsurance companies). The drive to differentiate appears to have been an important factor in risk financing innovation.[13]

*Convergence and globalization of financial services*

Broad regulatory and legal changes in financial institutions have led to a degree of integration between banking, investment banking, securities and insurance sectors.[14] The operational integration of financial institutions has led to a significant degree of interaction between formerly separate areas of financial services. 'Balance sheet' risk management, whereby financial institutions sought to connect a range of traditional banking, investment and insurance products to create an integrated range of financial products for clients, is one of the more notable illustrations of convergence influencing product design, selection and marketing.

Possibly less direct than other factors, it still may be argued that the rapid globalization of financial markets influenced risk financing innovation in the sense that globalization created a greater free-flow of capital and these more easily accessed resources provided building blocks for innovation.

### Capital market pressures

Globalization and convergence have directly affected capital markets, but it is worth noting that other influences have led to growing capital markets. Privatization of social security and pension systems around the world has released funds that initially had to compete for a rather limited range of investment opportunities.[15] Changes in regulations have further reduced frictional costs, and the immense size of the global capital market has pressurized financial institutions to create new opportunities for investment.

It is worth mentioning that the difficulties experienced in financial markets in the most recent three years have also directly contributed to a slowdown in innovation as banks and investors have grown more cautious.

## Can we classify and explain the most recent approaches to innovative risk financing?

### Risk transfer, retention and neutralization: foundation concepts

At the most basic level, risk financing can be categorized by the means (how will risks be financed?) and timing (when will the financing arrangement be triggered?) of the financing mechanism.

### How financing occurs

A firm finances risks directly or arranges for some other party to finance risk – risk is retained or risk is transferred. These alternatives are not discrete or mutually exclusive, and thus there is an intermediate category where retention and transfer are both present. One specific intermediate category is neutralization.

*Risk retention* refers to methods of self-financing. The risk is borne by the organization and if risk-related costs arise, the organization directly bears those costs. *Risk transfer* refers to methods of transferring the cost of risk (or more properly, the direct financial responsibility for the risk) from one party to a non-related party. Insurance is the obvious illustration.

Within the continuum of measures that fall between retention and transfer a wide range of tools reveal combinations of retention and transfer features. A simple example is an insurance contract with a deductible.

*Neutralization* is a specific category of retention/transfer combinations whereby a balance is specifically and purposely struck to neutralize the effects of risk (both down- and upside effects). The introduction of neutralization into the basic framework is significant because it permits the inclusion of (1) financial risk tools used for strictly financial risk purposes and (2) financial risk tools used for a wider array of risks.

### When financing occurs

The second dimension relates to the timing of the financing approach. The financing mechanism could be triggered before, during or after resources are needed. More formally, certain financing tools are *prospective* and involve arrangements to accumulate or secure resources in advance of a particular need. Other tools are *contemporaneous*, meaning the financing of a risk occurs as the resources are needed (indeed, often a risk event triggers the contemporaneous mechanism). Yet other tools are *retrospective* in the sense that they are structured to provide financing after events have occurred giving rise to a need for resources (see Figure 7.1).

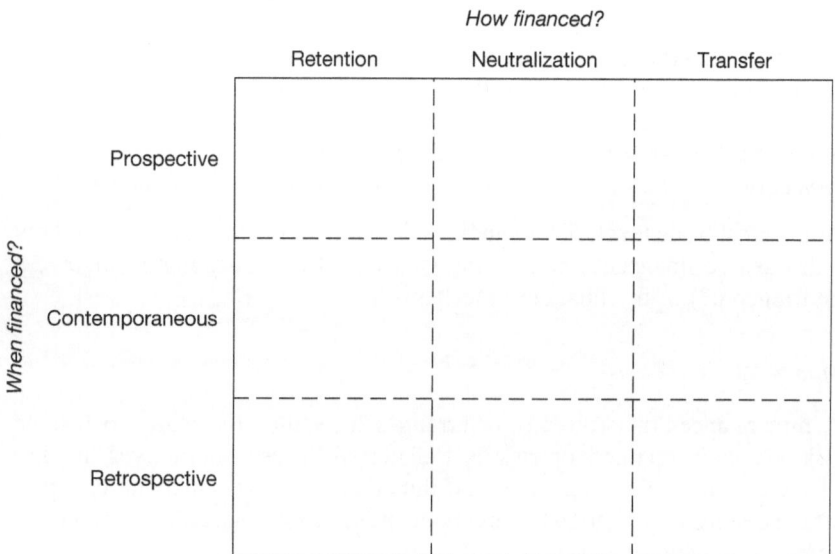

*Figure 7.1* The risk financing framework

Source: Young and Tippins[16]

Although practical issues can arise over the question of 'when' – and they will be discussed below – it is the 'how' question that evokes the most serious discussion today. And, central to that discussion is the question, what makes a transfer a transfer?

The courts and regulators tend to define risk transfer by (1) the legal validity of the arrangement, (2) the nature of the consideration of the parties, and (3) the pooling of risks. Superficially, the first point seems self-explanatory; transfer of risk must be legal. However, on closer inspection, the definition of a legal transfer requires that the risk be fully transferred outside the organizational boundaries of the transferor. Moving a risk from one member of a corporate family to another is *not* transfer in the eyes of most courts today (there are some occasional exceptions).

The second point, the nature of the consideration, also is significant. The transferor of risk is generally expected to make an economic commitment to pay for the benefits of transfer. Some courts have further noted that the commitment (e.g. a premium payment) must show some relationship to the risk to be transferred, although this is not a consistent requirement. The transferee (acceptor of risk) offers consideration in the form of an aleatory and conditional promise – that is, a promise to perform (e.g. pay a claim), conditional on some further chance event such as a loss.

The third point, the presence of pooling, is important. Traditionally, pooling referred to the notion of a risk bearer accepting many similar risks and pooling those risks together to accumulate funds sufficient to cover expected claims as well as fluctuations from expected levels of claims. Recently, some revision in thinking has been made regarding what is actually happening as a pool of risks grows. While historically the view has been that it is the actual accumulation of risks that leads to stabilization, a newer proposal suggests that it is the accumulation of funds – not risks – that is the logic behind pooling.[17]

This may appear to be a purely academic matter, but the changing view has very important implications. Traditionally, risk transference was closely associated with insurance and this meant that the distinction between retention and transfer could be (and was) based on evidence of a pool of risks. Is a risk-accepting party offering a similar arrangement to many other parties and are those accepted risks being pooled together for financial purposes? The newer way of looking at transference requires only that a pool of resources be available to cover the promises made under the contract (e.g. capital markets).

Unfortunately, regulation and the law have not fully kept pace with changes, and though it will not be discussed further here, it is important to know that risk financing tools and innovations are subject to regulatory and legal frameworks that do not match current realities. Many arrangements, otherwise completely indistinguishable, are treated differently by regulators solely based upon who is selling such a product (e.g. a bank or an insurance company). Indeed, one part of Attorney General Elliot Spitzer's recent investigation of the insurance and broking industries hinges on the inappropriate use of special purpose vehicles to facilitate finite risk insurance programmes (a 'traditional' innovation). These programmes have, in some instances, been found to not transfer risk (as their structure seems to imply) but simply to hide liabilities and risks.[18]

### Innovation building blocks: futures, forwards and options

Historically, risk financing innovations have tended to occur on the transfer or retention sides of the framework. Insurance firms experimented with variations on risk transfer (loss portfolio transfers, paid loss retros), while risk managers experimented with different retention models (pools, risk retention groups, captives, finite risk reinsurance). These older innovations will be discussed briefly in the final section of this chapter, but are covered more fully in Chapter 6.

On the financial risk management side, the use of futures, forward and options contracts enabled firms to address fundamental financial risks. Capital market-based risk financing innovations only emerged in the 1990s when experimentation began to occur to see if financial risk management tools might be used to manage a broader range of risks.

A *futures* contract is an arrangement for securing the future purchase or sale of goods or services at a predetermined price. *Forward* contracts do much the same thing but tend to be customized to specific buyers and sellers, whereas futures contracts are traded in markets and can be bought and sold anonymously. The purpose of a futures contract is to allow the buyer of that contract to neutralize the risk associated with some particular asset that the buyer will be acquiring or selling at some future point in time.[19]

An *options* contract is one in which the holder has a right to buy or sell some asset at a specified time and amount. In this case, however, an options contract may or may not be exercised (futures transactions must occur at the agreed date).[20] From a slightly different perspective, one might say that options contracts and insurance contracts have a great deal in common. If a firm buys an insurance policy, it has purchased an option that it may choose to exercise (by filing a claim when a loss occurs) if it is in the firm's interest to do so.

As will be seen below, more conventional financial instruments, such as bonds and certain equity contracts, also have become part of the capital market-based innovation toolkit.

### Capital market-based innovations

Futures, forwards and options could be implemented to address many financial risks. However, it is their application to risks that are not –

strictly speaking – financial risks that is the starting point for most innovations that have occurred over the past decade. Furthermore, these current innovations address somewhat deeper issues, such as capital structure; that is the balance between a firm's equity and debt. Therefore, as these tools will come to be seen as either 'equity-' or 'debt-' based arrangements, the following background comments are helpful.

### Capital structure considerations

Risk costs are significantly influenced by a firm's leverage. Such costs, transactional costs and bankruptcy costs for example, increase – not surprisingly – as risk increases for a firm. The inverse also is true. If a firm manages its risk, the costs of risk lower and the negative effects are mitigated. In general terms we would say that neutralizing (or hedging) risk would, therefore, add value to the firm.

Risk financing might be characterized as, in part, 'leverage management', which is to say risk financing measures are developed to assure that future possible events do not negatively affect the balance of debt and equity in a firm's capital structure. Establishing arrangements in advance of adverse events offers several general benefits including reduced problems related to possible bankruptcy, costs imposed by creditors and others, and possible liquidity, regulatory and tax implications. The obvious goal is to develop arrangements that allow a firm to seek new funding from capital markets in the best possible terms.

Leverage management may involve strategies that utilize capital market tools, and these will be the subject of further discussion below. However, leverage management also can include such things as dividend policy, investment decisions utilizing internal funds and, in some instances, even the legal form of organization (say, converting an insurance company from a mutual status to proprietary legal form).[21]

### Timing issues in innovative risk financing

If an event occurs (currency exchange movement, or a fire to a warehouse) requiring new funds, simple post-event risk financing might involve either issuing new equities, or acquiring debt. Frequently, such situations are of sufficiently modest impact that no prior arrangements need be made. The firm initiates the process retrospectively (post-event). Frequently enough some events may have such a negative impact on the firm that it either (1) cannot raise adequate funds, or (2) the costs and terms are punitive. Securing financing arrangements prospectively can address part of the problem (costs and terms), but the availability of funds is only one issue. The other is the effect of new funds on leverage. If debt is the source of funds, the effect on capital structure might lead to other problems for the firm.[22]

In any event, the issue of prospective arrangements is not clear-cut. It seems possible that markets could identify a firm with a high susceptibility to large events (or with financial vulnerability to the consequences of such events) *prior* to any adverse events occurring. Although there are exceptions to this point, in general it is not entirely clear that a firm would get any particular advantage in a prospectively arranged financing strategy because the market would build that vulnerability into the cost of the funding arrangement. Perhaps it is safe to say here that the actual advantage of prospective funding is not as great as might appear to be the case. Research has shown that the only clear, obvious and consistent benefit of pre-arrangements is fund timing. If access to funds is heavily time-sensitive, pre-arrangement may be critical.[23]

An important issue embedded in the question of financing timing is the matter of re-investment. It is not always a foregone conclusion that post-event re-investment, or even new investment, is in the best interest of the firm. Certainly, if the issue is, say, liquidity, the case is more easily made. Thus, if the firm holds high levels of illiquid assets, a financing arrangement can convert assets from illiquid to liquid without harming firm value. However, there are circumstances when a firm may not retain sufficient value to justify re-investment after a major event (and this illustrates how a specific event, in and of itself, can change the value of a firm).[24]

### Prospective 'contingent' arrangements for financing

As was suggested above, pre-arranging funding for post-event needs *may* be an improvement on pure post-event financing. The firm is not held completely victim to circumstances and the market. A line of credit (LOC) is, perhaps, the simplest possible example of prospective 'contingent' financing. The cost to the firm is the commitment fee and possibly the level of the interest rate.

An LOC does offer the benefit of producing funds quickly when needed, but previously the concern was raised that new debt can negatively affect the capital structure of the firm. Thus, the firm might be more inclined to consider *equity* contingent financing. There seem to be four general ways this could occur.

First, the firm could issue a put option on its own stock with a set striking price. These options would permit the firm to issue new stock. So, for example, a catastrophic event that depressed the firm's stock price would put the firm in a position to exercise the option. Since presumably the post-event stock price would be lower than the option striking price, the firm obtains equity funds for re-investment. Looked at slightly differently, we could say that the difference between the post-event stock price and the striking price is 'partial insurance'.[25]

A second possible approach would be a 'double trigger' put option.[26] Rather than basing the option trigger on a striking price alone, there is a further necessary condition (an event). The risk financing innovation here is that double trigger options have moved from the confines of purely financial risks (interest rates, currency exchange, credit) to include classically insurable events like catastrophic fires, law suits and other adverse events arising from non-financial perils.

The intriguing appeal of double trigger put options is that they span the known world of options and link with the newer world of insurable risks and thus serve as an experimental bridge between the two worlds. Pragmatically, this has meant that investment banks are more willing to experiment with such products and buyers are more comfortable with the form and structure. Further, these tools obviously link events with a broader measure of impact (stock price) and therefore seem consistent with an ERM conceptualization of risk financing.

Notably, double trigger put options are used by a number of insurance firms to partly insure underwriting portfolios against catastrophic loss experience. These options – known as CatEPuts (Catastrophe Equity Put Options) – may eventually rival traditional reinsurance products. CatEPuts are but one capital market approach used in the insurance industry today.[27]

Alternatively, an insurance company places a catastrophic risk (designated as one that pierces a predetermined level of accumulated losses) directly with investors. Such tools are most commonly seen (though they are not widely adopted anywhere yet) with natural catastrophes such as Californian and Japanese earthquakes, and Florida hurricane risks. Significantly, the evolution of these tools is viewed as an outgrowth of a more familiar – though indirect – approach, which is for investors simply to hold shares in a reinsurance company. The distinction here is that these equities are tied to a specific risk line or peril, not to the insurer's entire business. But it is also due to this conceptual connection that this approach is sometimes referred to as *securitization*.

Relevant to the discussion of innovation as a phenomenon, it should be noted that the greatest degree of risk financing innovation to date has, indeed, occurred in what are called the catastrophic (cat) markets. Analysts note that, unlike areas where innovation has only slowly advanced, cat markets seem to have a set of 'necessary ingredients' to assure success. These are:

1  a clear and direct demand for new capacity;
2  available technology and know-how to create new financial instruments;
3  statistical data in sufficient abundance to develop pricing models;
4  on-hand financial resources to invest in R&D;
5  reasonably savvy buyers and sellers.[28]

A third possible approach is 'hedge-bundling' as a customized strategy for financing firm risks.[29] Forwards, futures or basic options contracts can be packaged together (even with traditional insurance coverage), either in independent and unrelated contracts or as integrated multiple trigger-style contracts, and these arrangements can allow a careful selection of risks against which the package is structured.

Hedging, in principle, can be shown to be value adding to a firm, but the hedge-bundling approach has some important limitations. It can be costly and time-consuming, and the newness of this approach has led to reluctance in the market to accept it. On the other hand, it does show how traditional financial risk tools are being deployed for different purposes.

Fourth, since debt financing presents potential capital structure issues, an alternative might be to find the means of converting debt to equity after an event. Such a strategy would serve the dual purposes of providing funds for investment, but not affecting the firm's leverage. A rather straight-forward way to do this would be to buy back equity and issue new stock, though this does not assure that firm value remains unaffected. Another possibility that is seen with financial risks is the placement ('embedding') of a conversion option in the existing debt of the firm. Unlike most such arrangements, where the bondholder has conversion rights, these options are convertible by the firm. Although these arrangements can be compli-cated to understand, the process whereby a firm exchanges lower value stock for higher value debt also is a form of (at least) partial insurance against loss. Such financial instruments have existed for some time, so the relevant point is that such tools are consciously evaluated as strategies for dealing with non-finance risks (like insured perils). These instruments, known as 'reverse convertible debt' are not widely seen in the risk finan-cing world yet, but have properties similar to double trigger put options and seem to be likely candidates for further innovation.[30]

### Capital market innovation: a mirror-image illustration

One area where some advancement has occurred is in weather-related financial products.[31] The interest among some firms and governmental entities has been particularly high because these are arrangements that compensate organizations for economic impacts related to a variety of weather-related phenomena – excess snowfall, drought, extended periods of low or high temperatures. Traditional insurance does not address weather-related losses unless they produce direct physical damage, but weather can affect budgets, disrupt revenue streams and produce myriad effects.

A good illustration of this innovation is the risk of excessive/ inadequate snowfall. Imagine a large private resort catering to winter recreation enthusiasts (downhill and cross-country skiing, snowboarding, trekking, snowmobiling). For a business like this, the presence of snow is central to success. Therefore, the principal weather risk for this firm is inadequate snowfall. However, a lesser-but-significant risk might be exces-sive snowfall, inasmuch as the budget for maintaining roads in the winter would be strained and – after a certain point – even too much snow could curtail business.

TOO MUCH SNOW

Large resorts estimate costs for snow removal and, on average, they tend to be reasonably accurate. However, on a year-to-year basis wild swings in costs can occur. Since eliminating budget surprises is desirable, a risk manager might wonder whether it is possible to protect the snow removal budget from the risk of excess loss (meaning greater snowfall than is budgeted for). Ideally, the risk manager might like to arrange a risk financing contract that would produce additional funds to support snow removal if the existing budget is imperiled by unusual snowfall amounts. How might this happen?

One solution is a 'weather option'. This would be a contract that would tie payout to an agreed schedule (so many dollars per excessive inch of snowfall), that serves as the option trigger. The premium and the payout are both tied to snowfall rather than to loss, so in this sense the resort could receive payouts with little in the way of claims investigation and negotiation. If the resort is a publicly traded firm, a striking price trigger based on stock valuation might also be present, and thus this option has (or could have) the characteristics of the double trigger options discussed above.

The first stage of designing a weather option would be to quantify in financial terms the resort's exposure to excessive snowfall. For example, the resort may indicate that it budgets for 100 inches of snowfall between October and April of each year. Further analysis may reveal that cumulative snowfalls of up to 130 inches would not excessively tax the resort's budgets. So the realm of concern would be those years in which snowfall exceeds 130 inches.

In such a scenario, other elements will influence the determination of the contract price. Certainly 30 inches of snow in one storm has a different impact than 30 one-inch snowstorms, so time concentration may be a variable. Ambient temperature can also influence the impact of a particular snowfall. Heavy snowfalls can occur with temperatures near 30 degrees, but snowmelt occurs quickly, whereas snow falling in subzero temperatures may remain on the ground for weeks.

In any event, the resort will want to understand the level of snowfall it is willing to tolerate and the likelihood that this level will be exceeded. Further, a measurement of the impact of excessive snowfall will occur ('for every one inch of excessive snowfall, it costs our resort an additional $15,000').

The purported benefits of this customized option over conventional insurance include:

- these arrangements can be set for multiple years at a time;
- since most payouts are set by official statistics, payouts occur quickly with little difficulty;
- insurance taxes are not applicable;

- unlike insurance, the resort will almost invariably receive a benefit from the financial arrangement;
- certain revenue flow benefits may occur, even when no losses occur;
- these contracts can deal with numerous risks that traditional insurance does not.

TOO LITTLE SNOW

A second generic approach to weather risks is a weather bond. Like mortgage-backed bonds or securities, the resort could issue a bond whose proceeds might be tied to snowfall rather than to interest rates. Various triggers might be imagined here. Obviously, snowfall below some designated value could be the triggering event, but it could also be geared to drops in revenue, or to accumulated costs in artificial snow production.

There is more than one way to set up weather bonds. Conceivably, the firm could simply issue the bond directly as a kind of low-grade bond and hold the accumulated funds in reserve to cover costs resulting from too little snowfall. As a type of low-grade bond, these instruments would present a default risk that is not present at the same level in traditional bonds – the principal is not guaranteed – but equally, the return paid would be higher.

There are problems with this approach, but the most obvious issue is the matter of leverage. Adding additional debt to the firm may be problematic for capital structure. Recall that prospective funding arrangements typically make sense when funding is a time-sensitive issue. It is not clear that the impact of reduced snowfall is necessarily time-sensitive.

In order to address the leverage issue and, indeed, even the time-sensitivity issue, an alternative bond arrangement has been the creation of a special purpose vehicle (SPV) that would serve as the issuer and manager of bond funds. A sufficient arms-length relationship would exist so as to avoid – at least in concept – the harm to capital structure. Thus, a firm could have funds available to address snow shortfalls, but *largely* avoid leveraging issues.

The word 'largely' is emphasized above since the use of SPVs has come under scrutiny in recent years with their abuse in many high-profile corporate scandals (Enron being the most well known). In such instances, the SPV was used to hide debt off the balance sheet, and – as would be true with a weather bond – this can produce a misleading picture of the firm's leverage. Indeed, as has been mentioned previously, The New York Attorney General, Elliot Spitzer, recently has begun an investigation to determine whether such vehicles are, in fact, legal. However, for all intents and purposes, the current legal challenges should be understood as focusing on corporate intent. The fact that SPVs have been used for conceivably illegal purposes does not mean that properly configured vehicles cannot meet legal and managerial expectations.

In both illustrations, the key point to note is that financial risk management tools are being restructured and reconfigured to meet a wider range of business risks. And, this appears to be the essence of the most recent wave of innovation in risk financing, the utilization of financial management tools to more comprehensively address business risks – and importantly, using these tools to assure that both funds are available to meet post-event needs and that the funds meet capital structure and firm value objectives of the organization.

## What are the issues, challenges and problems that attach to innovations in capital market-based risk financing?

The actual degree to which capital market-based innovation has occurred in risk financing is limited (see, for example, Figure 7.2). The reasons for this are complex. Efforts to develop new tools in the mid- to late 1990s were dramatically curtailed by the downturn in the global economy in 2000, and have remained suppressed by events since then. Even the seeming recovery since 2003 has not brought about much renewed interest in risk financing innovations. Innovations in the late 1990s were driven by many factors but the clear champions for change were investment banks, reinsurance firms and insurance brokerage houses. Each has had common and distinct business challenges since 2000 (many of which are pressing and serious), and so enthusiasm remains low for new innovations.[32]

However, it must be said that, while external factors interrupted momentum, other factors may also play a role in the slowness of the innovation process. Infrastructure costs for new financial innovations are high (R&D, market formation, acquisition of technical knowledge, legal and regulatory issues) and these barriers to market formation have meant that early innovations have been costly – prohibitively so in some cases.[33] Within that range of barriers, specific suspicions exist over whether markets for these innovations can be adequately developed and maintained. To be fair, the turn in the market robbed analysts of the chance to learn that answer, but there have been concerns from the very early days of innovation that theory and reality might not neatly join together.

The technical complexity of many innovations, especially those involving hedging-type contracts has been another issue cited by risk managers. The mathematics of these products often are beyond the understanding of even highly educated managers, and they appear (to some risk managers anyway) to introduce a new type of financial risk that changes a firm's overall risk profile. Add to this perception the dramatic examples stretching from Orange County, California to Enron to Parmalat and beyond, and there is arguably sufficient uncertainty about risk financing innovation to keep the brakes applied in the near future.

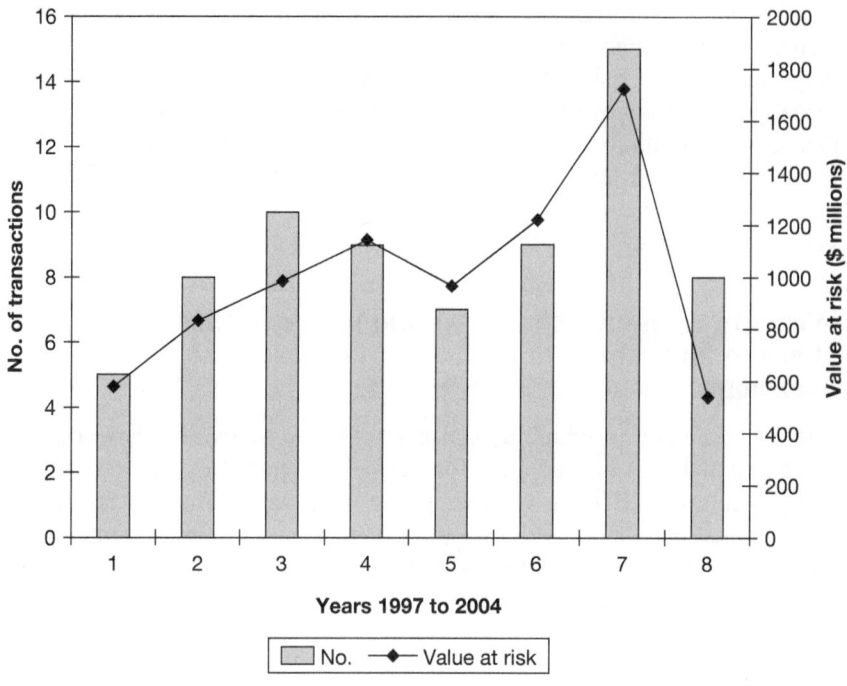

*Figure 7.2* Catastrophe bond transactions
Source: Charles Taylor Consulting[34]

Still, one might reasonably wonder whether, had the economy remained positive between 2000 and 2004, innovators might have worked through many of these issues. Certainly, as is true with most innovations, the transactional or unit costs could have diminished as the volume of business grew. And, uncertainty among buyers might have subsided somewhat as risk financing innovations became a more familiar feature of the risk management world. But, of course, at this point in time we can only speculate.

Indirect evidence exists to permit speculation on the preceding point. As Table 7.1 shows, the major deals that have been undertaken since the mid-1990s have tended to take place in financial institutions, and especially in insurance companies. Plausibly, since the risks of concern were technical underwriting risks, one could speculate that specific financial and managerial expertise existed to evaluate and judge the value of these innovations. In this light, it also is possible to argue that, in addition to technical expertise, the motive to address these risks was intense as the risks were core business risks. Centrality of risk, as well as expertise, may be a factor in the practical utility of these most recent innovations.

*Table 7.1* A representative sample of innovative risk financing arrangements

| | |
|---|---|
| Nationwide Mutual | US$400 million in contingent surplus notes |
| Arkwright Mutual | US$100 million in contingent surplus notes |
| AIG | US$10 million in catastrophe linked bonds |
| USAA | US$500 million in catastrophe linked bonds |
| Hannover Re | US$100 million in portfolio linked swaps |
| St Paul Companies | US$68.5 million in loss linked notes and preference shares |
| RLI Corp. | US$50 million in catastrophe equity puts |
| Winterthur Insurance | US$290 million in catastrophe bonds |
| Reliance | US$40 million in catastrophe bonds |

Source: Press Reports[35]

The general rule for evaluating risk financing tools and strategies is to undertake those measures that maximize the value of the firm. This seemingly straightforward decision rule masks a complicated theoretical justification that cannot be addressed here, other than to offer the following observations.

First, a number of real world influences intrude on the analysis of risk financing innovations. Tax treatment is quite different depending on whether a tool is deemed to be insurance or not. Accounting treatment can be quite different from country to country (when is an event recognizable as a loss or a gain?). Even general business customs and practices can affect the comparable value of risk financing tools.[36] The structure of management within a firm may actually limit the ability of risk financing measures to be integrated in any meaningful way. Understanding these factors is, needless to say, critical to analysis.

Second, the value of particular tools will vary depending on circumstances. A classic issue is the valuation of post-event financing measures. Comparing, say, firm value before a large loss and after a large loss is difficult because the large loss can affect the valuation in obvious and not so obvious ways.

Third, it often is difficult to evaluate risk financing alternatives simply on the basis of financial theory criteria. While it is true that maximizing shareholder wealth is a desirable objective, on a case-by-case basis other stakeholder interests can intrude. Most obviously, regulators can insist that certain measures (insurance, for example) be employed even when a strict financial assessment would suggest that other measures are superior. Vendors, key customers, labor unions, surrounding communities and others can exert influence over the means by which risk financing is undertaken.

## Concluding comments

Recent risk financing innovations – at least thus far – seem to be the result of an interplay of many factors. It is difficult to prioritize these factors by degree of influence, though it seems fair to believe that insurance intermediaries and investment banks have served as innovation drivers. The advancement of the enterprise risk management concept also has played an important role, inasmuch as it has created a climate in which innovation has been sought, permitted and, in some cases, necessary.

Nevertheless, any summarizing comment on risk financing innovation probably should include the following observations:

1   It does not seem likely that any new outbreak of capital market-based innovation will occur in the short term. Basic business problems for all types of financial institutions have suppressed the interest in risk financing innovations.

2   While Point 1 is arguably true, the need for alternatives to traditional risk financing arrangements remains high. Insurance markets are unstable and, in some lines, highly unreliable. Since about 2000, efforts to experiment and solve insurability problems have been limited to 'first generation' innovations such as self-insurance pools, proprietary mutuals, captive insurance arrangements and conventional banking arrangements (finite risk insurance).

3   The general momentum in the risk management field will probably produce three long-term results:

   a   increasing sophistication and professionalization of the field;
   b   a more integrated outlook on the role of risk management within overall organization management;
   c   a view of risk financing that is increasingly in line with broader financial management principles.

As a result, demand-side pressure for innovation is likely to grow, though perhaps at a slow but steady rate.

More generally, the experience thus far with risk financing innovation suggests (but only suggests) general issues seen in any type of innovation. Hurdle costs are high, and they tend to discourage early adopters. The uncertainty associated with technical complexity tends to limit adoption too, and this may be particularly acute in fields where the management of risk and uncertainty are central functions. While risk taking has entered the lexicon of risk management, it remains rather difficult to make the case that a well-risk managed firm should be aggressively innovating in the means by which risks are managed.

Other general observations include the issue of intellectual or professional receptivity to new ideas. In some sense, buyers need to be able to clearly articulate their needs for innovation – even if they cannot articulate the nature of the possible innovation itself. To date, the advocates for innovation have been mainly (not only, but mainly) brokers, investment banks and insurers. Risk managers and their firms have not yet taken full ownership of the demand for innovation, though changes in the field of risk management arguably could lead to that result.

It is worth noting in passing here that 'traditional innovations' have continued to flourish (see Figure 7.3), and this suggests – at least indirectly – several of the above points. As the figure indicates, alternatives to insurance have grown consistently for the last two decades, which is interesting for two reasons; first, alternative financing arrangements now are nearly equal to the size of the commercial insurance market and, second, the alternative market did not shrink at any point during the extended soft insurance market of the 1990s (in other words, once a firm decides to opt out of the commercial insurance market, it tends to not return). Clearly, frustration with the unpredictability of insurance markets is continuing to drive risk managers into alternative risk financing strategies. These tools and techniques are discussed more fully in Chapter 6 but, as a point of contrast, it is useful to note that while capital market-based innovations have not re-emerged in recent years, the impulse to innovate is present.

Finally, an issue largely unaddressed here is the matter of innovation boundaries. Almost by definition, the concept of innovation is boundary-free, but in the case of risk financing, there may be practical limits to the application of capital market strategies to traditionally insurable risks. Researchers have shown over time that pure and speculative risks have different properties, and that – importantly – people perceive and react differently to risks that have possible positive outcomes (called speculative

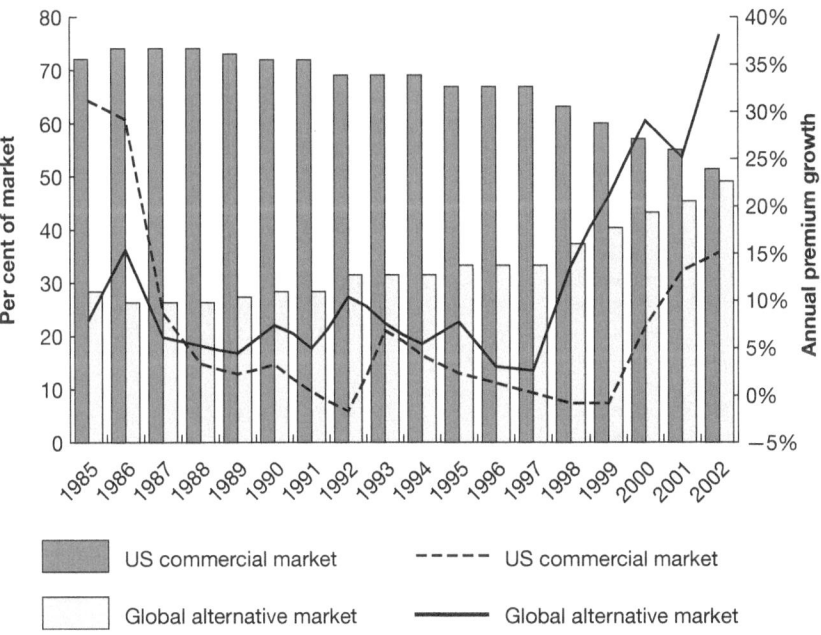

*Figure 7.3* Global utilization of alternative risk transfer, 1985–2002

Source: Charles Taylor Consulting[37]

risks) and those that don't (pure risks).[38] Notably, the issue of moral hazard becomes critical but in different ways. In speculative risk settings, risk taking is encouraged and the properties of such risks are such that many people will want to capitalize on incentives. The concern in such settings is to keep behaviour legal. In pure risk settings, moral hazard can collapse almost any risk financing stratagem, and the goal is to purge the risk financing programme of moral hazard properties. It remains to be seen whether the moral hazard issue is, in fact, a boundary between the application of certain risk financing innovations and certain risks, but in theory, it might be.

## Notes

1  William, C.A., Jr, M.L. Smith and P.C. Young, *Risk Management and Insurance*, 8th Edition, New York, NY: Irwin McGraw-Hill, 1998, Chapter 2, covers a fair bit of the history of risk management.
2  Ibid.
3  Hull, J., *Options, Futures and Other Derivative Securities*, Englewood Cliffs, NJ: Prentice Hall, 1993 provides a meaningful introduction to the subject.
4  William, C.A., Jr, M.L. Smith and P.C. Young, *Risk Management and Insurance*, 8th Edition, New York, NY: Irwin McGraw-Hill, 1998, Chapter 1.
5  Hogan, W.P., 'Corporate governance: lessons from Barings', *Abacus*, Vol. 33, no. 1, 1997 provides a cogent assessment of the risk issues arising from the Barings scandal.
6  The Committee of Sponsoring Organizations of the Treadway Commission, *Enterprise Risk Management – Integrated Framework*, Jersey City, NJ: AICPA, 2004, Chapter 3.
7  Ibid.
8  William, C.A., Jr, M.L. Smith and P.C. Young, *Risk Management and Insurance*, 8th Edition, New York, NY: Irwin McGraw-Hill, 1998, Chapter 13 covers the topic of risk financing and provides an overall view of the topic.
9  Lamm, R.M., 'The catastrophe reinsurance market: economic gyrations and innovations amid major structural transformation', *Insurance and Weather Derivatives: From Exotic Options to Exotic Underlyings*, edited by H. German, London, UK: Risk Books, 1999.
10  Beck, M., L. Drennan and P.C. Young, 'From Cadbury to Turnbull: finding a place for risk management', *Proceedings of the Third International Conference on Money, Investment and Risk*, Djangoly Innovation Centre for Europe, Nottingham Trent University, 2004.
11  Ibid.
12  Standards Australia, *AS/NZS 4360: Risk Management Standard*, Sydney, Australia: Standards Australia, 1999. This standard has been updated in 2004 but was not available at the time this chapter was prepared. This standard generally is recognized as having effectively encapsulated the major practitioner-driven developments in the field.
13  Young, P.C. and S.C. Tippins, *Managing Business Risk*, New York, NY: AMACOM Books, 2000, Chapter 6.
14  International Financial Services London, *International Financial Markets in the UK*, London, UK: IFSL, November 2003, Sections 2 and 3.
15  Young, P.C. and S.C. Tippins, *Managing Business Risk*, New York, NY: AMACOM Books, 2000, Chapter 6.
16  Ibid.
17  Ibid.

18   Ibid.
19   Doherty, N.A., *Integrated Risk Management: Techniques and Strategies for Reducing Risk*, New York, NY: McGraw-Hill, 2000, Chapter 6. This book is an overall excellent resource on the relationship of risk management to financial management in firms.
20   Ibid.
21   Froot, K.A. and J.C. Stein, 'Risk management, capital budgeting, and capital structure policy for financial institutions: an integrated approach', *Journal of Financial Economics*, 47: 55–82, 1998. A well-known and respected article on the relationship of risk management to capital structure.
22   Campbell, T.S. and W.A. Krakaw, 'Corporate risk management and the incentive effects of debt', *Journal of Finance*, 45: 1673–86, 1990. One of several notable articles that examine the role of debt management in risk management strategies.
23   Froot, K.A., D. Scharfstein and J.C. Stein, 'Risk management: co-ordinating investment and financing problems', *Journal of Finance*, 48: 1629–58, 1993.
24   Doherty, N.A., *Integrated Risk Management: Techniques and Strategies for Reducing Risk*, New York, NY: McGraw-Hill, 2000, Chapter 9.
25   Ibid.
26   Froot, K.A. and J.C. Stein, 'Risk management, capital budgeting, and capital structure policy for financial institutions: an integrated approach', *Journal of Financial Economics*, 47: 55–82, 1998.
27   Doherty, N.A., *Integrated Risk Management: Techniques and Strategies for Reducing Risk*, New York, NY: McGraw-Hill, 2000, Chapter 12.
28   Ibid., Chapter 16.
29   Ibid., Chapter 14.
30   Stein, J.C., 'Convertible bonds as backdoor equity financing', *Journal of Financial Economics*, 32: 3–21, 1992. This article is technically oriented but still presents the subject in a reasonably straightforward manner.
31   Ramamurtie, S., 'Weather derivatives and hedging weather risks', in *Insurance and Weather Derivatives: From Exotic Options to Exotic Underlyings*, edited by H. German, London, UK: Risk Books, 1999. A well-acknowledged introduction to the subject.
32   International Financial Services London, *International Financial Markets in the UK*, London, UK: IFSL, November 2003, Sections 2 and 3.
33   Doherty, N.A., *Integrated Risk Management: Techniques and Strategies for Reducing Risk*, New York, NY: McGraw-Hill, 2000, Chapter 16.
34   From a presentation by Martin Fone of Charles Taylor Consulting, 11 January 2005. Charles Taylor Consulting is a London-based alternative risk financing consultancy.
35   Lamm, R.M., 'The catastrophe reinsurance market: economic gyrations and innovations amid major structural transformation', in *Insurance and Weather Derivatives: From Exotic Options to Exotic Underlyings*, edited by H. German, London, UK: Risk Books, 1999.
36   Young, P.C. and S.C. Tippins, *Managing Business Risk*, New York, NY: AMACOM Books, 2000, Chapter 6.
37   From a presentation by Martin Fone of Charles Taylor Consulting, 11 January, 2005.
38   Young, P.C. and S.C. Tippins, *Managing Business Risk*, New York, NY: AMACOM Books, 2000, Chapter 6.

# Reference

Young, P.C. and S.C. Tippins, *Managing Business Risk*, New York, NY: AMACOM Books, 2000.

# 8 The evolution of enterprise risk management

*Gerry Dickinson*

## Introduction

During the last decade, risk management within enterprises (i.e. companies and other organizations) has recently been broadening its scope. It has evolved into a more systematic and integrated approach to the management of the total risks that an organization faces. This development can be traced to two main causes. First, following a number of high-profile company failures and preventable large losses, the scope of corporate governance has widened to embrace all the significant risks that an enterprise assumes. Directors are now increasingly required to report on their internal risk control and compliance systems. This is either through voluntary codes, such as the Combined Code of the UK Listing Authority, or by legislation, as in Germany through the 'Control and Transparency in Entities' Law. In the US, the Sarbanes-Oxley Act (2002) and the new Enterprise Risk Management Framework issued by COSO (the Committee of Sponsoring Organisations of the Treadway Commission), published in 2004, extends the scope of corporate governance even more widely.[1] The second influence on the development of enterprise risk management has been the greater role that shareholder value models have been playing in strategic planning. Early strategic planning models paid insufficient attention to corporate risk. Modern strategic planning models are now based more on shareholder value concepts, which draw their inspiration from finance theory where risk has always played a central role.

## Origins of enterprise risk management

Risk management as a formal part of the decision-making processes within enterprises is traceable to the late 1940s and early 1950s.[2] There were two earlier, if limited, strands of risk management practice that have more recently been integrated under the broader concept of enterprise risk management. One of these strands relates to the management of insurance risks and financial risks.

For many years, enterprises have been able to transfer certain types of risks to insurance companies. These transferred risks related to natural catastrophes, accidents, human error or fraud, but as the scope of insurance markets expanded, some types of commercial risks could be transferred, such as credit risks. The existence of these insurance markets forced managers to consider alternatives to the purchase of insurance. Some of these insurable risks could be prevented, or their impact reduced, through efficient loss prevention and control systems and some could be retained and financed within the company. This led to a broader approach to the management of insurable risks.

Since the late 1970s companies have been looking more closely at how they manage various financial risks that they face: currency risks, commodity price risks, interest rates and credit risks. Financial risk management began, as a formal system, at the same time as the development of financial derivative products: financial futures, options and swaps. This was no coincidence, since investment banks had developed these financial instruments and their associated markets in part to allow their corporate customers to hedge these financial risks. Hence, financial risk management emerged in much the same way as insurance risk management had previously. It was stimulated by the existence of these new financial instruments, which caused management to consider how much of the risks should be retained within the company and how much should be offset through these external arrangements. The availability of financial derivatives also forced enterprises to consider more carefully the pricing of risks, how risks could be financed internally, and to assess the value of the additional services supplied by investment banks.

Management also recognized that insurable risks and financial risks should be considered together, since the purchase of insurance and the purchase of derivatives to hedge financial risks performed essentially the same role. This recognition led in the 1990s to the development of new integrated risk transfer and risk securitization solutions that combine both types of risk.

The second strand in the development of a more holistic approach to risk management arose from more general management thinking. Contingency planning had been a part of corporate policy for many years, its purpose being to identify those activities that might be threatened by adverse events and to have systems in place to cope with these events. Business continuation management extended the practice of contingency planning by requiring more comprehensive internal systems. Both contingency planning and business continuation management approaches, however, were limited, since they presupposed that strategic choices had already been made and their role was confined to the effective implementation of these strategies. Enterprise risk management has extended these approaches further so that risk management is now an integral part of corporate strategy formulation as well as implementation.[3]

## Defining enterprise risk

Enterprise risk is the extent to which the outcomes from the corporate strategy of an enterprise may differ from those specified in its corporate objectives, or the extent to which it fails to meet these objectives (using a 'downside risk' measure).[4] This clearly presupposes that these corporate objectives are realistic targets, and that they are not too ambitious or too easy to achieve. The strategy selected to achieve these corporate object-ives embodies a certain risk profile, which derives from the factors that can be expected to impact on the activities, processes and resources chosen to implement the strategy (see Figure 8.1).

A wide range of external and internal factors can cause the outcomes from a corporate strategy to depart from those set down in its corporate objectives. Some external factors relate to those in the marketplace in which an enterprise operates, such as new entrants into the market, changing consumer tastes or new product developments. Other external factors arise from a wider context, such as changes in the economy, changes in capital and financial market conditions, and changes in the political, legal, technological, demographic and other environments. Most of these are beyond management control, although active risk management requires that there are systems in place to make an enterprise more resilient and adaptable to major changes. Risk management is a dynamic process.

Another set of factors that can cause outcomes to differ from those planned arise from within the enterprise itself: human error, fraud, systems failure, the disruption of production and so on. These are increasingly referred to as 'operational risks'. Clearly, management has a greater degree of control over loss producing events that take place within the organization compared to those that arise outside.

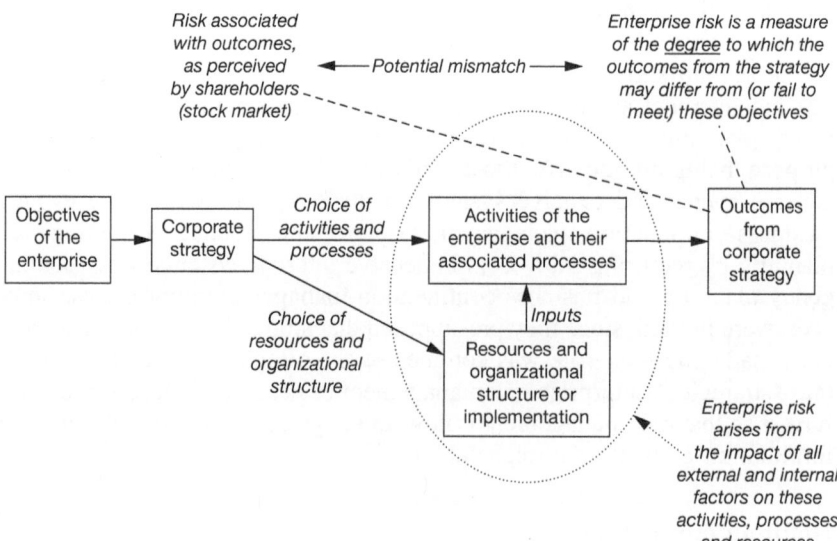

*Figure 8.1* Measuring enterprise risk

In seeking to assess the impact of a very wide range of external and internal factors on the activities of an enterprise, there must be some simplification to make the task manageable. With the assistance of computer modelling, scenario analysis is increasingly being used to analyse and measure the joint impact of external and internal causes of risk on the enterprise over a planning period. Risk mapping is also being used in order to prioritize the risks that the enterprise faces, so that the risks that could have the greatest impact are given the most management attention.

If enterprise risk is measured in terms of corporate objectives, a consistent framework of analysis is produced. But there are also shareholder value models to consider. Shareholder value models specify that the corporate objectives of a company should be coincident with those of shareholders. However, shareholder risk can only be determined indirectly, since it depends on how the stock market perceives and values the riskiness (volatility) of the expected net cash flows (future earning streams) from a company's activities. When the corporate objectives of a company are aligned with those of shareholders, enterprise risk will be close to the overall risk of the company as perceived in the stock market. But it should be kept in mind that in both competitive product markets and in risk-averse stock markets, a corporate strategy with a higher risk profile will tend to have higher rewards: there is a compensation for the lack of predictability.

## Retaining and transferring risk

Since the overall risks of an enterprise are an integral part of its corporate strategy, one way of managing these risks is through the choice of the corporate strategy itself. If top management consider the risk profile of a particular strategy to be too high, it can change the strategy to one with a lower risk profile. Hence, enterprise risk management must be a top-down process, but with an efficient system of information feed-back.

Just as other corporate decision-making processes take place in a hierarchical structure, so do risk management decisions. The questions of whether to buy insurance or to hedge financial risks depend on the strategic decisions that have already been made. For example, the currency risks of a company arise *because* it has international activities. Thus, if a company's production is located in a country with a strong currency relative to those countries it wishes to export to, one way of managing these currency risks is by relocating its production facilities.

Most of the risks that an enterprise faces cannot be insured or hedged, and so they must be retained and financed internally. Other mechanisms also exist for reducing risk, apart from the purchase of insurance and hedging with financial derivatives. Legal mechanisms are one of these, as some risks from commercial activities can be restricted through the use of corporate vehicles by exploiting their limited liability status. Large-scale projects and major real estate developments are often structured this way. Similarly, most risk securitization arrangements exploit the risk reduction benefits of limited liability through the use of special purpose vehicles.

Divestment of corporate activities and the outsourcing of operating functions provide other mechanisms for risk transfer. But unlike insurance, hedging or legal mechanisms, divestment or outsourcing represent a transfer of a commercial activity itself, and not just the risks embedded in these activities.

Decisions on how much insurance should be bought, how much of the financial risks that are faced should be hedged, or the degree of divestment and outsourcing that might take place, will be largely determined by a few key considerations. The scale of potential loss or, more precisely, the greater the potential adverse impact on the attainment of corporate objectives, the greater will be the management's preference for risk transfer rather than risk retention. Decisions on the balance between risk retention and risk transfer will not just be related to their scale of impact. The degree of information and competence that the enterprise possesses in managing a specific set of risks are also important.

When an enterprise divests or outsources an operation, it usually does so because it considers the recipient to be better equipped and more knowledgeable in managing these activities. For example, the outsourcing of computer and information systems to a specialist organization can reduce the risks of technological obsolescence and systems failure, as well as increasing cost effectiveness. More knowledge and a greater core competence usually mean lower risk, since the impact of a risk event often depends on who is managing or controlling the underlying process. Similarly, in buying insurance or derivative contracts to hedge financial risks, information to assess the likelihood and severity of potential loss is an important part of the decision-making process. Enterprises will tend to have less information on the underlying probability distributions needed to price insurance risks than insurance and reinsurance companies, especially if the insurable events occur only very infrequently, such as 9/11 or the Asian tsunami. This also applies to the requisite information to price financial risks; investment banks tend to have more information and better financial engineering methodologies than corporate treasury departments.

The dynamic relationship between corporate strategy and risk acceptance (retention) and risk transfer/hedging decisions is outlined in Figure 8.2. If the risk/return profile of a potential corporate strategy is not in line with what management considers its shareholders and other key stakeholders would prefer, it can follow one of two major courses of action. It can retain its chosen strategy and seek to change its attendant risk/return profile through the purchase of insurance, hedging of financial risks, outsourcing some of the activities associated with the implementation of the strategy, and/or the use of corporate structures which exploit the risk reduction features of limited liability. Alternatively, management can change the risk/

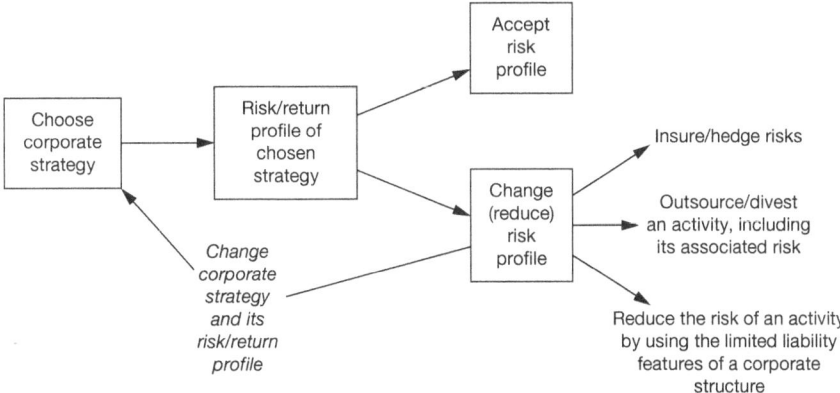

*Figure 8.2* Strategy, risk retention and risk transfer

return profile by changing the corporate strategy itself. Because managing the risk through a change in corporate strategy is a high-level decision, risk management must be a top-down process to be fully effective.

## Enterprise risk and risk hierarchies

Much attention has been devoted in the theory and practice of enterprise risk management in analysing risks in terms of: (1) their underlying causes; (2) the likelihood and severity of their impact; and (3) the degree to which risk events are within the control of management or not. But there has been insufficient attention paid to the hierarchical nature of risks within an enterprise. The importance of recognizing the risk hierarchies arises from two main factors. First, risks are an integral part of an organization's decisions and activities. Second, corporate decisions and activities are themselves hierarchical in nature.[5] Strategic-level decisions give rise to a sequence of lower level planning and implementation decisions. For example, an enterprise may decide to pursue a greater degree of international diversification to meet its corporate objectives. Such a strategic decision will entail choosing a set of activities in a select number of countries. This strategic-level decision will, in turn, give rise to a sequence of products, production and distribution decisions. How these product, production and distribution processes are to be financed is an even lower level decision in the hierarchy. Within an efficient organization, the hierarchy of decisions will be considered jointly prior to the final commitment to a given strategy, but in many cases lower level decisions can only be realistically taken at a later stage when a strategy is being implemented.

In general, higher level decisions not only set the agenda for lower level decisions but they also determine their time sequencing. Hence, higher level decisions impose a potential constraint or conditionality on lower level decisions. Sound management practice will seek to reduce these constraints

by thinking through the details of the implementation of a strategic decision in advance and by putting in place effective information feed-back
and reporting mechanisms to reduce any significant costs from these
constraints. Thus enterprise risk management should be viewed as managing the whole hierarchy of risks: from higher level strategic risks through
to lower level risks, such as financial and operational risks. Greater
emphasis needs to be placed in designing an enterprise risk management
policy by mapping risks onto decision hierarchies. This will ensure a
greater consistency with strategic management and its implementation, of
which enterprise risk management forms an integral part.

## Some general propositions on enterprise risk management

This broader concept of enterprise risk management also gives a clearer
positioning on how insurable risks and treasury or financial risks should
be viewed within the organization. Insurable risks and financial risks are
both subsets of enterprise risk. Hence, if there were no insurance markets
and no derivative markets or other hedging mechanisms, all the risks that
an organization faces would be enterprise risks, since they arise as a
consequence of the activities that it undertakes.

We can summarize the above and our earlier discussion on enterprise
risk into the following propositions:

1   Enterprise risk is embodied within a corporate strategy (i.e. the choice
    of corporate activities and the choice of resources and organizational
    structure to implement these activities) which in turn is influenced by
    uncertain environments.
2   Enterprise risk can only be effectively *measured* in terms of corporate
    objectives. The *degree* of risk is the extent to which the actual outcomes
    from the activities of an enterprise differs from (a variance concept
    of risk) or fails to meet these corporate objectives (a 'downside'
    concept of risk).
3   Where the enterprise is a quoted company, the more closely aligned
    the corporate objectives set by management are to the preferences of
    its shareholders, the closer will be the enterprise risk to the stock
    market's own risk assessment of the company.
4   Since the financing of the risks should be integrated into the overall
    financing of the enterprise, insurance buying and self-insurance deci-
    sions and hedging policies need to be closely coordinated with the
    wider cash management and capital structure decisions.
5   Risk retention decisions on insurable risks (e.g. choice of deductible
    levels) and risk retention decisions for financial risks (e.g. choice of
    'strike prices' on option contracts) should be determined jointly; both
    types of risk are subsets of the overall enterprise risk and hence
    correlations between them need to be considered.

## Enterprise risk management and creating shareholder value

For companies, especially companies listed on stock markets, a key issue is how to relate enterprise risk to shareholder risk. It was noted above that when the corporate objectives of a company are aligned with those of its shareholders, enterprise risk will be close to the risk that is perceived by its shareholders. This risk will not be exactly the same as that perceived by shareholders, since company management can only know approximately the risk perceptions and risk preferences of its shareholders, or indeed the collective view of all existing and potential shareholders within the wider capital market. But by defining corporate objectives close to what management think the preferences of their shareholders are, this mismatch will be reduced. Hence, management is much more likely to reduce this mismatch when its overriding corporate objective is to earn a rate of return on its equity capital that is at least as high as the rate of return that its shareholders could earn on alternative investments for the same level of capital market risk (i.e. the cost of equity capital).

The dynamics of risk management must also be considered. Capital market theories often assume that the corporate strategies of companies can be captured within a risk-return framework, where the overall risk (enterprise risk) is considered to be mainly outside the control of management.[6] A corporate strategy is chosen by a company within a set of environments and it is these environments that determine the risk-return profile of the strategy. In other words, management can choose a strategy from a number of feasible strategies, each having a different risk-return profile, but the expected risk-returns on these different strategies are largely outside the control of management. This is true to a large degree, as management cannot influence changes in the economic, socio-political, technological, commercial or other external environments that will largely determine the degree to which its corporate objectives are fulfilled. In the same way, the captain of a ship cannot control the sea and weather conditions that could occur on a particular journey and hence ensure that the ship arrives at its destination on time. In practice, however, some risks are within the control of management, especially those arising from human actions within the enterprise itself, including operational risks. Management can reduce some of these risks in advance or can reduce their impact if an adverse event should arise. Just as a good ship's captain can take action during a journey to avoid hazards that will slow down the journey, and indeed prevent the ship from sinking. This ability of company management to reduce risk is not adequately recognized in capital market theory. Thus, a sound enterprise risk management system can reduce shareholder risk without reducing expected shareholder returns, providing the costs associated with risk prevention and containment are not too high.[7] When this occurs, shareholder value is created. This gain to shareholders is depicted in Figure 8.3 in terms of a shift in the efficient investment frontier that they face.

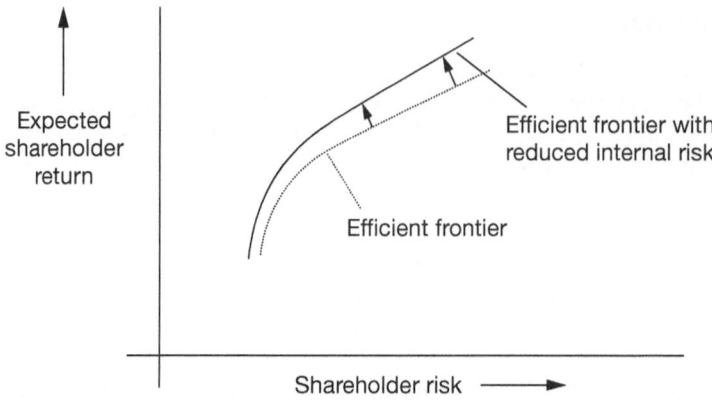

*Figure 8.3* Improvement in the efficient investment frontier facing shareholders

## Organizing the function of enterprise risk management

How should enterprise risk management fit within an organizational structure? Since enterprise risk management must be a top-down process, the chief executive and the senior executive team should determine the parameters for the policies and the organizational structure for its effective implementation. At the same time, there must be information feed-back from those closest to the sources of risk, so that senior managers are well informed when formulating their overall risk policy. In addition, management must delegate some responsibility to those closest to where the risks are likely to impact so that early action can be taken to prevent a small problem growing into a larger one.

Because of the complexity of identifying, controlling and managing risks across an enterprise, dedicated and specialist expertise is required. A new coordinating management role is now emerging – that of the Chief Risk Officer (CRO).[8] The CRO, who is usually a senior executive and part of the top strategic planning team, may retain a more traditional job title, such as Group Risk Director, even if his/her responsibilities have now widened, but the title of Chief Risk Officer is growing in use.

In addition, the CRO must maintain close links with the Chief Financial Officer (CFO). The financing of risks, whether retained or transferred, rests with the CFO, who will inevitably be a senior executive and will also sit on the main strategic planning committee. The CFO is responsible not only for the purchase of insurance and derivatives, since these decisions fall within corporate treasury function, but also for the overall financial policy of the company, which includes the financing of all retained risks.

Corporate governance standards now require boards of directors to develop more clearly defined risk audit functions, including an overview of their top management teams.[9] This high-level risk audit function is often an additional responsibility for the audit committee of the board of directors. Since executive directors themselves have to be monitored, a non-executive director should chair the audit committee in order to give it the necessary degree of independence. The board of directors has the ultimate responsibility for the enterprise risk of the company, being accountable to shareholders and other stakeholders. In countries where there is a practice of having a dual-board structure – an executive board and a supervisory board – the CRO should report to the supervisory board. The structure of reporting, risk policy guidelines and information flows for an efficient organization is depicted in Figure 8.4. In practice, the CRO often reports to the CEO as he/she is a senior member of the top executive team, but it is important from a corporate governance standpoint that there is also an independent reporting line to the non-executive directors or the supervisory board.

Recently, there has been a discernible increase in the number of appointments of CROs in banks, insurance companies and other financial services firms. This increase has been prompted not just by corporate governance concerns but by changes in government regulatory and supervisory systems. Following the lead of the Basel Committee on Banking

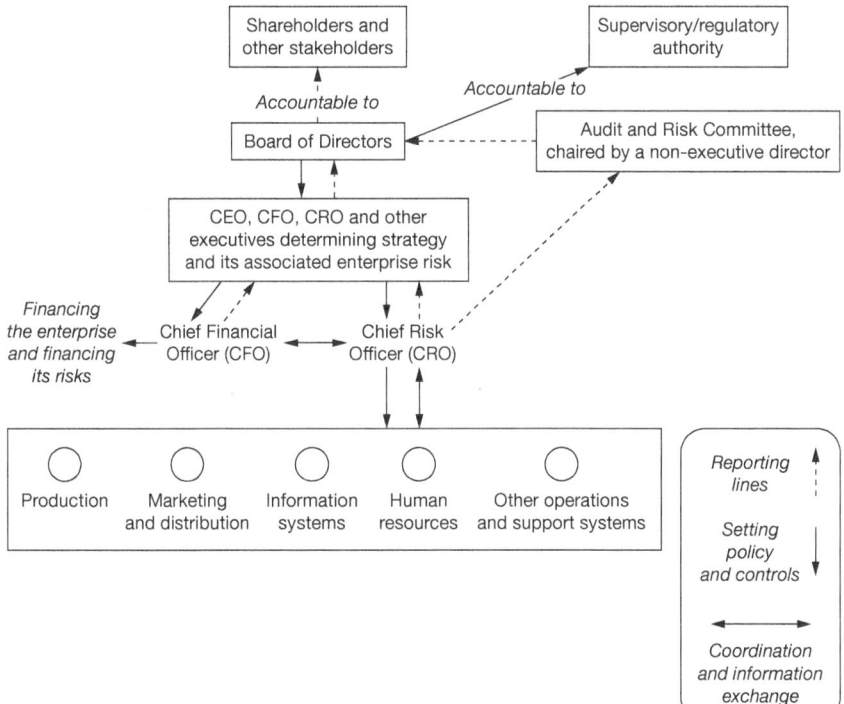

*Figure 8.4* Enterprise risk management and its organizational setting
(for company with a unitary board of directors)

Supervision under Basel II,[10] national regulators and supervisors are requiring that directors of banks, insurance companies and securities firms have effective internal risk management systems. The International Association of Insurance Supervisors (IAIS) and the International Organisation of Securities Commissions (IOSCO) have been active in supporting this initiative by the Basel Committee. Internal risk management systems, including asset-liability risk management models, are seen by governments as complementing prudential regulation and supervision systems. Governments are increasingly providing incentives to financial services enterprises to have sound and transparent risk management systems in place by granting them less stringent solvency and capital adequacy benchmarks.[11] While the ultimate responsibility for the risk management must rest with the board of directors, the CRO is now being given not only the overall functional responsibility for the enterprise risk management system but also some liaison role with the supervisory authorities.

## Conclusions

Enterprise risk management has been strengthening its position within the strategic planning process over the last decade. Moreover, under the new corporate governance environment, boards of directors and top managers are now being made more directly accountable for the risks that their enterprises assume and so more financial resources and the top-level support will be forthcoming to help the role to develop further. For financial services firms, recent changes in regulatory and supervisory regimes, arising from the three-pillar approach inspired by Basel II, are giving enterprise risk management a further stimulus. The challenge will be to find the individuals with the right blend of skills to assume these enhanced responsibilities. Individuals will have to possess a good understanding of corporate strategy, finance, law, and the complex operating processes within the organization, as well as an ability to communicate well, both inside and outside the enterprise.

## Notes

1  PriceWaterhouseCoopers, 2004, *COSO Enterprise Risk Management – an Integrated Framework*. New Jersey: American Institute of Certified Public Accountants.
2  Kloman, H.F., 1992, 'Rethinking Risk Management' *The Geneva Papers on Risk and Insurance*, No. 64, July, pp. 299–313; Giairini, O. and Stahel, W.R., 1993, *Limits to Certainty: Facing Risks in the New Service Economy*, 2nd revised edition. New York: Kluwer Academic Publishing, pp. 54–60; and Dickinson, G.M., 1989, 'Corporate Risk Management in the Age of Global Networks' in *Strategic Trends in Services: An Inquiry into the Global Service Economy*, ed. A. Bressand and K. Nicolaidis. A Services World Forum Project. New York: Harper & Row, Ballinger Division, pp. 237–251.

3   For a discussion of the widening scope of risk management see: Dickinson, G.M., 1997, 'Integrating Insurance and Hedging into the Overall Risk Management of the Firm' *Singapore International Insurance and Actuarial Journal*, Inaugural Issue, Vol. 1, August, pp. 161–173; and Young, P.C. and S.C. Tippins, 2001, *Managing Business Risk: An Organization-Wide Approach to Risk Management*. New York: AMACOM Press, pp. 4–14.

4   Dickinson, G.M., 2000, 'Risk Role Grows to Enterprise Scale' *Mastering Management, Part 7*. London: *Financial Times*, 13 November, Supplement, pp. 14–15. Reprinted in *Mastering Management Vol. 2*, ed. James Pickford, 2001, FT-Prentice Hall, pp. 191–195. Essert, H., 2002, 'Risk and Enterprise Value' *The Geneva Papers on Risk and Insurance*, Vol. 27, No. 3, July, pp. 435–443.

5   Grant, R.M., 2002, *Contemporary Strategic Analysis: Concepts, Techniques and Applications*, 4th edition. Oxford: Blackwell, pp. 197–202.

6   Brodie, Z. and R.C. Merton, 1999, *Finance*. London: Pearson Publishing, pp. 267–298.

7   Doherty, N.A., 2000, *Integrated Risk Management: Techniques and Strategies for Managing Corporate Risk*. New York: McGraw Hill, pp. 225–229.

8   Lam, J., 2003, *Enterprise Risk Management: From Incentives to Controls*. New Jersey: J. Wiley & Sons, pp. 43–51.

9   Deloach, J. and N. Temple, 2000, *Enterprise Wide Risk Management: Strategies for Linking Risk and Opportunities*. London: Financial Times Management, pp. 30–35.

10  Bank of International Settlements, 2001, *Proposal for a New Basel Capital Accord*. Basel: Bank of International Settlements, January; and Bank of International Settlements, 2004, *Basel II: International Convergence of Capital Measurement and Capital Standards: a Revised Framework*, June.

11  Ross, S.A., 2001, 'Financial Regulation in the New Millennium' *The Geneva Papers on Risk and Insurance*, Vol. 26, No. 1, January, pp. 8–16.

# 9 Intellectual property and bridging loans

## Their emerging roles in venture finance and business rehabilitation in Japan

*Masatoshi Kuratomi*

This chapter provides a glimpse of emerging business models within Japanese financial institutions, most of which suffered many problems because of the large number of non-performing loans (NPLs) in existence since the 1990s. The NPL problem reduced the funding opportunities for new industries and technologies in Japan and drove most Japanese banks into management difficulties. The Development Bank of Japan (DBJ)[1] a governmental institution, together with private financial institutions, has taken some risks in order to provide a variety of new financial structures to support and enhance the current transition, especially in the two following sectors: venture finance and business rehabilitation.

## DBJ's venture finance programmes

### Targets of venture finance

Venture finance in Japan is still in its infancy, as traditional financial institutions, such as banks, have been based heavily on finance schemes collateralized by tangible assets, especially land or estates. The decline in the economy from the early 1990s and the deflation[2] experienced since the late 1990s have forced land prices to spiral downward, and the traditional Japanese banking style known as 'land collateral finance' has been reviewed, with new business models in the banking sector now being planned and implemented.

DBJ has assisted and promoted these changes from the standpoint of revitalization of regional economies through the proper assessment of new businesses. In cooperation with a number of regional banks, DBJ had proposed some special financial measures for venture businesses depending on their operational stages.

DBJ identifies four stages for targeting venture finance. First, incubation funds, which develop utility performance based on theoretical or laboratory findings and/or the acquisition of patents, have become operational recently on a regional basis, for example Tohoku Incubation and Kyushu

Venture Partners. These funds, which deal with start-ups to early-stage ventures, have been capitalized by regional investors and DBJ. Second, measures for financing early-stage to middle-stage ventures include providing loans or guarantees secured by warrants. This form of finance scheme is available for ventures in which the value of intellectual property is difficult to assess but shows great promise. The third way of financing middle-stage ventures is by providing a loan or guarantee collateralized by intellectual property. This is a very effective way for most ventures that lack tangible assets as ordinary collateral to finance development expenses. Finally, loans or guarantees secured by tangible assets are available for later-stage ventures in the same way as for other businesses.

There are also opportunities for the application for loans collateralized by receivables or movables for ventures at all stages, as an additional security measure.

DBJ's purposes behind financial assistance for business ventures can be divided into four areas: quantitative supply for a venture that has a reasonable track record, but for which funding for development expenses is unobtainable because of lack of traditional collateral; qualitative assistance in the long term, for example for three to five years, and fixed-rate finance complementary to funds available through private banks; assistance to strengthen corporate governance by project-finance evaluation, and to accelerate business motivation in techno-ventures by providing fair evaluation of their intellectual property; and assistance to strengthen creditworthy status through credit analysis by a governmental bank, and to improve public awareness through mass-media publication.

### Intellectual property as collateral

Loans or guarantees collateralized by intellectual property are ways to meet future R&D expense needs through the evaluation of business opportunities derived from such intangible assets. The category includes two kinds of evaluation: of intellectual property in developing a pipeline to finance its future business development; and of developed intellectual property to finance new R&D expenses. Both types of intellectual property should be evaluated and collateralized, but what conditions are necessary for this?

Availability for quantitative evaluation is one of the essential elements. Intellectual property is given a present value through a discount method based on projected cash-flow; therefore, it is noticeable that the technical advantage is less important than with current cash-flow; potential or future value should not be taken into account and intellectual property that does not produce any cash-flow under evaluation will not be available as collateral.

Another essential element is security as collateral. Liquidity and transferability are important for a transferee to continue conducting business with the intellectual property. Security on all the necessary intellectual property rights for operating a business is also essential. Patents and/or copyrights that provide a business venture with core competence should be taken as main collateral, while related design rights, trademarks and manuals may be taken as supplementary collateral.

Intellectual property in Japan is categorized as follows: software copyrights are based on Copyright Law, and can be registered at SOFTIC.[3] Their period of protection will run from just after their creation to 50 years after the death of the author. Other copyrights, such as music, pictures and databases, can be registered at the Agency for Cultural Affairs and are protected in the same way.

Industrial intellectual property, such as patents, designs, trademarks, trade names, goods indications and integrated circuit plans, are registered mainly at the Japan Patent Office and will be protected for between six and 20 years from the time of application or registration.

Methodologies to evaluate intellectual property are based on discount cash-flow calculations for five years at the most. Projected cash-flow at present value may be considered as fair collateral after careful analysis of sales figures calculated by unit price and the number of units, life span, version-up development expenses and its timing. The discount rate for present value calculation depends on track record, market risk or worn-out risk, but currently discount rates of between 10 and 20 per cent are widely applied throughout Japan. Transfer costs, i.e. initial expenditure for human resources and for fixture transfer, are also not to be forgotten and should be deducted for final assessments.

Great attention should be paid to collateral rights protection, otherwise smooth transfer may be readily eroded. The most important point is to clarify the relationship between the multiple patents related to one product. As for the relationship between a fundamental patent and a manufacturing or product patent, there is no allowance for producing a specific product only through using a fundamental patent and it is possible to offend fundamental patent territory with only a manufacturing or product patent. It is also imperative to pay attention to third-party rights, such as licensing and joint development. If rent or execution rights are necessary for business, expenditure for those rights should be considered as an additional costing.

Registered intellectual property is not enough for continuing business as before. Customer databases, factory plans and special fixtures and equipment should be carefully kept to avoid dispersion, and even unregistered software, databases and manuals should be copied for transfer. Pivotal workforces in the existing staff should be kept employed after a transfer as (or because) a specific technology has been developed and integrated with (those) specific human resources.

Having provided over 200 loans[4] collateralized by intellectual property in the last nine years, DBJ's experience shows that the reality of execution of collateral transfer in the event of business failure has been rather difficult so far. Intellectual property owned by troubled enterprises tends to become worthless as salvage value, unless the reason for the failure is not based on the product itself but on a funding gap stemming solely from companies' operations. A liquid market has not been developed for intellectual property, even though Techno-mart, a Japanese market for unused intellectual property, works cooperatively with companies. Bilateral talks between the related counterparts, areas of which have been naturally limited, have been a main, practical solution. Therefore, lenders should, as a precaution, prepare internal credit alarm systems as an effective solution to seeking possible M&A opportunities before the business troubles of borrowers become apparent.

### Loan structures secured by warrants

Warrant finance is a kind of ordinary loan agreement combined with subscription agreements of warrant: loans are provided in exchange for warrant allocations.[5] DBJ may obtain payback of a loan by selling these original allocations to management, or their business counterparts, in the latter stages without warrant execution in order to obtain any shares, or by selling them to the market when the IPO[6] is made.

This scheme may apply to those ventures whose intellectual property cannot be collateralized but whose business plans are expected to be successful. That finance is available for those who do not have any collateralized property is not the only advantage of the scheme; in addition, applicable interest rates can be set lower than those of an ordinary loan and a careful warrant allocation scheme can provide current shareholders with an anti-dilution effect.

On the other hand, share dilution is considered disadvantageous and the debt amount repayable will increase under this scheme in line with simple capital increases. Warrant finance cannot be a main source of venture finance in this regard, but combined finance from venture capital can work favourably as an anti-dilution measure.

From a banking point of view, profit from a sold warrant should fulfil a future default risk within a warrant finance scheme account. A warrant is not a direct form of collateral and needs portfolio management based on a considerable number of loan executions. If you evaluate increased warrant value properly, reduced interest rates can be applicable in some specific cases. But the sale of a warrant just after loan execution cannot be made because profitable warrant portfolios need to be kept in reserve for future defaults.

. According to the Black-Scholes Model, a higher share price will lead to the greater value of a warrant. This also applies when higher volatility in a share price, a lower price of warrant execution, a longer time to maturity, and a higher rate of interest are taken into account. Warrant value for venture businesses is rather difficult to calculate because of too much volatility, but a price tag of 1 to 5 per cent of the share price is often the reality. The original value of a warrant is the surplus obtained when the warrant is executed and can be explained as the difference between current price and execution price. The other factor is a time value that provides expected profit to maturity, which will reduce to none when matured.

There are several points for lenders to consider regarding warrant finance. Commercial Law requires resolution at a general meeting of shareholders for interest-bearing issuance, such as a warrant offer. A company that acquires a warrant should prepare to pay tax against expected profit when the warrant executed. It is often negotiated whether a discounted interest rate can be regarded as an acquisition cost. There is little need to pay attention to the Law of Maximum Interest Rate and the Law of Subscription concerning expected surplus, but the Banking Act and Monopoly Prohibition Law may be taken into consideration for securing liquidity.

### Receivables and movables

Receivables are available as a finance measure for business ventures. Those of a third party who has a higher credit rating would be useful as collateral, and using such receivables is easier and more favourable than a venture using its own credit. A claim agreement that includes the third-party debtor, the cause of the receivables, the contracting term and the maximum amount between lenders and borrowers should be prepared and needs to be registered. A related transfer notice or consent to/from a third-party debtor is necessary for secured collateral. This transfer notice to a third party raised the credit doubt of borrowers and was used in a restrained way until the mid-1990s, but, since there has been recognition of this kind of receivable finance scheme for balance sheet improvement, doubts have been dispelled.

Movables in warehouses are also available for finance in the same manner as receivables. A movables transfer agreement for collateral use should be prepared and a clear indication, such as a signboard or label on the transferred collateral, is necessary. It is important to understand that collateralized movables need enough liquidity for sale and detailed warehouse management for security. Registration schemes are under discussion for much easier finance. Movables finance is available not only for business ventures but also for business rehabilitation.

### SPC finance to create an animated TV programme

It is rather difficult to finance developing intellectual property, because there is no track record of cash-flow from the intellectual property and there may even be some uncertainty regarding its final registration as a legally protected right.

DBJ, together with The Bank of Tokyo Mitsubishi, has extended a loan to a new animation development through an SPC[7] finance scheme. Film and TV programme contents in Japan have been equity-financed by a sponsor group to date, which has provided individual group members with a rather complicated share of the intellectual property of the contents, and, consequently, most of the sponsors have been limited to contents-product professionals. The new scheme intends to provide integrated and effective management of the copyright on a long-term basis and to promote Japanese visual-contents industry development in introducing an opportunity for non-recourse loans and attracting a much broader base of investors to improve the financial infrastructure in this sector.

Gonzo Ltd, in charge of the animation development of the project, is one of the most advanced SMEs involved in two and three-dimension combined digital technology and its parent company, GDH Ltd, has been active in the copyright business in Japan.

DBJ established a new loan programme to promote effective use of intellectual property in the fiscal year 2004 and this has been the first case on the programme.

## DBJ business rehabilitation support programmes

During the Japanese economic downturn in the 1990s, many Japanese companies with sustainable and independent business capabilities and promising business opportunities faced difficulties in continuing because of overleveraged debts that have emanated from other business sectors within companies. These companies need to restructure their diversified businesses in order to decrease their debt burden by making a 'selection and concentration' to strengthen their core competence. Although a lot of work needs to be undertaken, troubled businesses or companies now have numerous opportunities to revitalize themselves through joining with external sponsors or through M&A activities.

### *Environment of business rehabilitation*

As described earlier, economic downturn and long-term deflation in Japan from the 1990s has led to asset deflation, which caused a huge number of NPLs at major banks or regional banks to decline in property-based collateral value.

But the recent situation concerning NPLs can also be observed in the reduction in the ratio of the number of NPLs to total loans outstanding. This ratio for major banks dropped steadily to 5.1 per cent at the end of the fiscal year 2003 from 7.1 per cent a year earlier. Regional banks showed a relatively larger fall to 6.9 per cent from 7.9 per cent in 2003. This was in contrast to developments during 2002, when the ratio for regional banks remained virtually level, while that for major banks declined.

Major banks' outstanding NPLs declined significantly in the fiscal year 2003, as they continued to remove large numbers from their balance sheets while a very small number of new NPLs emerged. At regional banks, NPLs that were removed considerably exceeded newly emerging NPLs, in contrast to previous years, when they had basically been equal. This result seems to indicate that regional banks have been becoming more aware of the need for sound management, given the public attention paid to their financial conditions and particularly to the ratio of disclosed NPL to total loans as an indicator of financial soundness. The ratios of some banks, however, remain high, particularly among regional banks, hence continued efforts are still needed to accelerate NPL disposal.

DBJ understands this situation and has prepared a special finance programme for business rehabilitation. It is joining with regional banks[8] to enhance their rehabilitation finance activities as well as the removal of their NPLs as a part of a strengthening strategy related to their Relationship Banking[9] activities.

### Rescue and bankruptcy procedures

Rescue and bankruptcy procedures in Japan consist of two frameworks, one of which is processed through legal administration that has different objectives: revitalization or termination. The revitalization process is consequently divided into three: a process under Civil Rehabilitation Law; a process under Corporate Reorganization Law; and a corporate consolidation by Commercial Law. The Civil Rehabilitation Law, which took effect in April 2000, has been providing successful outcomes[10] for troubled companies wishing to begin anew. This is due to the following characteristics: quick procedure (5 to 6 months for approval of the rehabilitation plan), less stringent conditions for creditors' approval of the rehabilitation plan, debtor-in-possession financing, available application for non-bankruptcy status and available creditors' option for the collection of the claim. The termination process in this framework consists of bankruptcy and special liquidation.

Private consolidation as the other framework for rescue and bankruptcy procedure is divided into several options and combinations of these: discharge of debt, deferred repayment, a process supported by Industrial Revitalization Law, and so forth. It is often agreed among stakeholders of SMEs as a private resolution, along with the Private Resolution Guidelines,[11] to identify its transparency and impartiality and to seek tax deduction against write-off as well.

### Finance programme for business rehabilitation

DBJ established a special finance programme[12] for troubled companies to bridge their required financial resources in the supplementary budget for the fiscal year 2001, a year before the first target year set by the government for the removal of NPLs from banks' balance sheets, as most private banks had not been familiar with, and had been cautious about, this risky banking area. This programme contributes to the government's measures for resolving the problems of NPLs at Japanese financial institutions and of overleveraged debt at enterprises. Under the new programme,[13] DBJ has been providing debtor-in-possession (DIP) financing for petitioned companies, acquisition financing[14] for business buyout by third parties and long-term loans for business restructuring in the latter stages of the rehabilitation process. DBJ also makes a number of equity inputs into corporate restructuring funds.

Businesses eligible for these programmes show commitment in the following areas: assurance of economic and social value and future development potential; confirmation of contribution to future community development and employment; confirmation of certainty of the rehabilitation plan; confirmation of repayment through sufficient security measures; confirmation of the agreed and appropriate revision of responsibilities among the stakeholders; and confirmation of the application of relevant due diligence.

DBJ's functions consist of four categories within a rehabilitation process. First, as the business rehabilitation plan is very important in achieving practical and improved performance, DBJ uses its experiences and networks to assist its planning in cooperation with lawyers, accountants, consultants, business sponsors, financial sponsors and creditors. Second, in the early stages of the process DBJ can provide a debtor with a short-term DIP loan for limited facility investment and/or working capital to keep its business value and continue its operation, through a prompt but detailed credit analysis. Third, after the approval of a rehabilitation plan, restructuring finance will be required on a long-term basis. Such refinance will be organized as a syndicated loan with private banks, but DBJ can also participate. Finally, DBJ can engage in equity participation in corporate rehabilitation funds in order to promote private investment. Those funds function as hands-on management as well as financial sponsors for restructuring companies.

Corporate rehabilitation finance is regarded as a type of structured scheme and its funding is provided as a syndicate loan in a similar way to a project finance scheme.

### *Corporate rehabilitation funds*

As of March 2004, DBJ resolved to invest in 21 restructuring funds, the total amount of which would be ¥530 billion. DBJ's investments are classified into three types: fund investments as a limited partner, co-investments in a fund exclusive to a specified company and equity investments in general partners. Outlined below are examples of DBJ's activities in this area.

### *Investment in Nippon Mirai Capital Co. Ltd*

DBJ undertook capital investment in Nippon Mirai Capital Co. Ltd. Upon its establishment in February 2002, DBJ contributed approximately

20 per cent of the new company's paid-in capital of ¥450 million, while the remaining 80 per cent came from the company's management members and private-sector business corporations.

This was DBJ's first investment in a corporate rehabilitation fund established to facilitate a comprehensive solution to financial and industrial restructuring. Nippon Mirai Capital is a management company for corporate rehabilitation funds. The scale of the fund will ultimately be between ¥50 billion and ¥100 billion, to which DBJ contributes as a limited partner as well. It has four characteristics that provide particular advantages: neutrality that is not categorized into any specific financial or industrial group, and which can therefore offer the best practice for revitalization; broad networks that are organized by the shareholders, the advisory board members and the external partners; provision of opportunity to financial institutions through accepting equity participation by post-DES[15] shares, which can provide the returns from future up-side performance; and the provision of opportunity to cautious Japanese investors to enhance risk money liquidity.

The fund will make investments only to those enterprises with high revitalizing potential and will analyse those investments on the prerequisite condition of a debt decrease by discharge or DES. The fund, as a means of investment, can buy the credits directly from lenders, such as banks, can buy or accept the equities converted from DES, and can buy shares in the rehabilitating company.

Lenders will discharge part of their claims and execute DES in order to secure opportunities to gain the up-side return of future revitalized performance. They will also be able to avoid their own shareholding issues derived from the shareholding regulations applied to banks and their monitoring work and holding cost, through contribution of the converted shares to the fund in exchange for a share of the limited partnership. The two options available on the converted shares, sale or contribution, can be chosen depending on the lenders' own perspective or tactics for the rehabilitation plan.

*Funds for specific companies – the Sakura-no Department Store (then DacVivre) Rehabilitation Fund*

In March 2002, DBJ resolved to invest in the DacVivre Rehabilitation Fund,[16] which was established to support the restructuring of DacVivre Co. Ltd, a regional retailer, in Sendai City, Miyagi Prefecture. The fund was established with capital contributions from companies in the Tohoku region, including Takeda, an estate company, and the Michinoku Bank,[17] together with those of the executives and employees and DBJ. This is the first rehabilitation fund established to support a specified company.

Although DacVivre initiated civil rehabilitation procedures in response to various adverse developments, including a chain reaction of business failures triggered by the Mycal[18] collapse, it is still able to generate substantial earnings from its core activities, and is considered to be a company that can be rebuilt through financial restructuring, business restructuring and other measures. Since July 2002, DacVivre has acquired new sponsors and altered its company name to Sakura-no Department Store after the approval of the Rehabilitation Fund in May of that year. DBJ, together with the Sumitomo Mitsui Banking Corporation, also extended a DIP loan collateralized by the receivables from credit card companies.

### Regional restructuring funds

DBJ has resolved to invest in the three funds that have focused on regional rehabilitation businesses to date: J. Wind Fund I and Renaissance Funds I and II. J. Wind Fund I is managed by J. Wind Partners and has been targeted to buy the NPLs extended to the SMEs with high rehabilitation potential through their balance sheet improvement. The total amount of the fund is expected to be ¥20 billion. Renaissance Funds I and II are managed by the Japanese subsidiary of BNP Paribas and these funds have also focused on regional SMEs. The total was closed at ¥14 billion for Fund I and is planned to close at ¥35 to 40 billion for Fund II.

DBJ has invested in these funds in order to contribute to regional economic revival and enhance regional banks in strengthening their Relationship Banking activities.

## Finance examples

### Exit finance from a process of Corporate Reorganization Law

Fuji High Polymer, a manufacturer of plastic housing materials, has been in a rehabilitation process according to Corporate Reorganization Law since 1988. As the company has produced regular earnings under the rehabilitation plan, a new financial structure with special covenants, including a disposition of unused land, was arranged and syndicate-funded[19] by Mizuho Bank, DBJ and other private banks, such as Nanto Bank,[20] in order to execute earlier debt repayment or provide exit finance.

This earlier debt repayment provided the debtor with management freedom from judicial supervision under the process and an increase in the company's credibility. It also provided the former creditors with the benefit of removal of their NPLs earlier than planned, while the private banks acted as a source of new DIP finance.

### Bills-receivable finance for a private rehabilitation process

Nippon Yakin Kogyo, one of the largest stainless steel producers, has been in a private rehabilitation process, working within the Private Resolution Guidelines, since 2002. Although most manufacturing companies in Japan depend heavily on their working capital funding being resourced from discounted bill receivables, the situation has been difficult for the financing of those companies, because the Guidelines do not obligate the lenders to maintain the discount funding balance during the rehabilitation process as they do for that of loans on bills, deeds and overdrafts.

The bill receivables of ¥8 billion owned by Nippon Yakin Kogyo were carefully separated from the company and trusted to Mizuho Trust Bank, and then divided into the senior and junior beneficiary rights. The senior part of the trusted rights has secured the loans as AA-rated, which provided the company with a low-interest funding opportunity from Mizuho Corporate Bank and DBJ during the rehabilitation period.

### Inventory finance

Peter Trading, an SME retailer of kids' clothing in Fukui City founded in 1914, has been in a rehabilitation process since August 2003. DBJ provided a first inventory-collateralized loan for ¥50 million as DIP finance, which enabled the company to repay a debt classified as an NPL at a regional bank.

As described earlier, DBJ has had some experience of the finance scheme collateralized by movables, but this has as yet to be widely recognized as a popular finance scheme due to the following two reasons: lack of a public registration system for movables and the unprepared secondary market for the liquidated property.

Prior to the establishment of the public registration system that is under discussion within the government, DBJ delegated the management of movables to Kiacon, a Japanese liquidation specialist allied with SB Capital Group[21] in the US, for the execution of inventory finance. Kiacon provides comprehensive or 'entrance to exit' management of movables, which consists of initial evaluation as collateral, advisory services and monitoring of monthly performance and execution of closing sales on liquidation.

Peter Trading will increase its credibility and trading volume and also obtain knowledge of inventory management through this new finance, which will motivate the company to achieve its rehabilitation plan.

### Finance for regional hotel business rehabilitation

Ochiai-ro, a countryside spa hotel in Izu Peninsula established in 1875, has been in a rehabilitation process since May 2002. This SME originally had two kinds of hotel: traditional Japanese 'ryokan'-style accommodation and a modern hotel focusing on tourist groups. The reason for the business failure was that the modern hotel had been in a slump after the bubble economy collapsed in the late 1990s. Also, the traditional hotel had lost its regular customers due to a deterioration in quality of service.

The Ochiai-ro Rehabilitation Fund that was established by DBJ and other investors in September 2003 became a new shareholder of the restructured company and has been active in the execution of the rehabilitation plan in conjunction with a loan from the Suruga Bank.[22] Under the scheme, only traditional accommodation is being planned due to its competitiveness. DBJ's support has been provided in respect of the traditional hotel's considerable influence on the regional economy and history.

### *Debt–debt swap*

Debt–debt swap (DDS) is a conversion of the existing NPL credit to the junior debt position. DDS has some varieties in terms of the types of conversions targeted: a long-term, bullet credit, in which interim repayments are interest-only repayments, the principal sum being repaid in the final bullet; a junior credit that can earn repayment only after the senior debt has been repaid; or a junior credit that can receive repayment only when the surplus cash-flow becomes obtainable. In general, DDS bears higher interest rates than senior loans because of the risks involved.

DDS will be able to gain more tax benefit then DES depending on debt status and profitabilty of a restructuring company. As the interest on converted debt by DDS can be deductible in order to reduce the taxable profit, the restructuring company stays profitable. On the other hand, dividends given to DES shareholders should be allocated from the final profit after the tax.

### *Cooperation between the Industrial Revitalization Corporation of Japan, the Resolution and Collection Corporation, private rehabilitation funds and DBJ*

The Industrial Revitalization Corporation of Japan (IRCJ) was established in April 2003 along with the newly enforced IRCJ Law, with the aim of providing revitalization assistance beneficial to both the industrial and financial sectors in Japan, through removal of NPLs from Japanese financial institutions. Within this role, IRCJ is seeking to assist companies that have sound business fundamentals but are unable to thrive independently because of excessive levels of debt and other factors.

The roles of IRCJ and private rehabilitation funds with DBJ's support can be misunderstood as duplicated or overlapped as governmental assistances in this sector. The IRCJ's main role is to arbitrate a rehabilitation planning process that has been deadlocked among private parties. Their decision is made from the public and fair perspective that is finalized by the Industrial Revitalization Committee,[23] which contributes complementarily to solutions for market problems. The equity of IRCJ was contributed to by the public[24] and private sectors.

The Resolution and Collection Corporation (RCC), another public body reorganized in April 1999 for the provision of solutions on the NPL problem and industrial revitalization in Japan, has prepared a trust scheme for SME corporate rehabilitation that is also complementary to the functions of private funds.

## Notes

1　Established in 1999 by the reorganization of two governmental institutions under the DBJ Law in order to contribute financially to Japan's economic and social policy by supplying long-term funds and related services to projects that promote the revitalization and sustainable development of the Japanese economy and society, a more affluent national life, and the independent development of regional economics. DBJ's services are intended to supplement and encourage those provided by commercial financial institutions.

2　GDP deflator and CPI have shown negative figures since 1994 (except 1997) and 1999 respectively. Cabinet Office and Ministry of Public Management, Home Affairs, Posts and Telecommunications. Available at: www.boj.or.jp/en/stat/stat_f.htm.

3　Software Information Centre. The competent authorities are the Ministry of Economics, Trade and Industry and the Ministry of Education, Culture, Sport, Science and Technology.

4　Since the fiscal year 1995, DBJ has provided 286 loans to business ventures that amounted to ¥16.3 billion, 230 of which were for ¥13.2 billion and were collateralized by intellectual property, as of March 2004.

5　Warrant is a common word for a security that offers the owner the right to subscribe for the ordinary shares of a company at a fixed date or price. However, it is usually allocated to DBJ as free or bonus shares in this context due to the insufficient present value of a business venture. This kind of warrant is often understood as 'para-collateral' leveraged by potential value of a business venture.

6　Initial Public Offering.

7　Special Purpose Company. This is commonly used as a vehicle for securing bankruptcy remoteness from sponsors in project-financed or asset-financed projects.

8　DBJ has alliance agreements with 83 regional banks as of June 2004.

9　The Financial Services Agency of Japan issued a policy paper.

10　The numbers of applications for the Civic Rehabilitation process are as follows: 3,635 cases for the last four fiscal years; 804 in 2000; 1,019 in 2001; 948 in 2002; and 864 in 2003. These cases occupy 70.1 per cent of the listed 77 bankruptcies, which are now regarded as a main legal procedure for listed companies.

11　'The guideline for multi-creditor out-of-court workouts' was produced in September 2001.

12　Loan and Investment Programme for Business Rehabilitation.

13　Loan approval amounted to ¥67.8 billion for 38 rehabilitations, as of March 2004.

14　The US Bankruptcy Code allows a troubled debtor to obtain credit through 'Chapter 11' in order to operate as an ongoing concern. In Japan, loan extensions for post-petition working capital under Civil Rehabilitation Law are categorized as DIP finance, but, in a broader sense, this can include financing for the post-approval stage of a rehabilitation plan.

15　Debt Equity Swap.

16　The fund will amount to from ¥1.2 to 2 billion.

17　A regional bank in Miyagi Prefecture.

18　Mycal was then the main shareholder of DacVivre and the fourth largest retailer in Japan. It collapsed in September 2001.

19　The total loan amounted to ¥1 billion.

20　A regional bank in Nara Prefecture.

21　SB Capital Group in the US contributed to the K-mart restructuring process as a liquidator.

22 A regional bank in Shizuoka Prefecture.
23 The Industrial Revitalization Committee is the authority established under the Industrial Revitalization Law to ensure fair and impartial decision-making at the IRCJ.
24 This public body is the Deposit Insurance Corporation of Japan (DICJ). DICJ is a semi-governmental organization that was established in 1971 with the purpose of operating the Japanese deposit insurance system, in line with the Deposit Insurance Law. The operations consist of four kinds of businesses: management of the deposit insurance system and inspection of financial institutions; financial revitalization against the failure of financial institutions; and operations related to sound financial institutions, such as capital injection; and collection of NPLs and pursuit of management liabilities of failed financial institutions.

# References

Bank of Japan (2004), *Overview of Japanese Banks: Observations from Financial Statements for Fiscal 2003*, July, BOJ Research Papers, available at: www.boj.or.jp/en/ronbun/ronbun_f.htm.

Financial Services Agency (2003), *Action Programme Concerning Enhancement of Relationship Banking Functions*, March, Opinion Paper by Relationship Banking Working Group, available at: www.fsa.go.jp/news/newse/e20030328-1a.pdf.

Japanese Bankers Association (2001), *The Guideline for Multi-creditor Out-of-court Workouts*, October, Private Resolution Guideline Working Group, available at: www.zenginkyo.or.jp/en/news/index.html.

Teikoku Databank (2004), *Trends of the Application for the Civil Rehabilitation Law in Four Years after the Law Was Put into Force*, April, TDB Watching, available at: www.tdb.co.jp/english/news_reports/w0404.html.

# Index